Nevil Maskelyne

Frontispiece. *Nevil Maskelyne*, aged about 44. Drawing in black and red chalks, on blue paper, attributed to John Russell who probably drew it about 1776. In the possession of Mrs H. C. Arnold-Forster. (Photo: NMM)

NEVIL MASKELYNE

The Seaman's Astronomer

DEREK HOWSE
With a Foreword by
Sir Francis Graham–Smith
Astronomer Royal

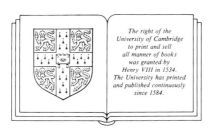

CAMBRIDGE UNIVERSITY PRESS

Cambridge

New York   New Rochelle

Melbourne   Sydney

Cambridge

Published by the Press Syndicate of the University of Cambridge
The Pitt Building, Trumpington Street, Cambridge CB2 1RP
32 East 57th Street, New York, NY 10022, USA
10 Stamford Road, Oakleigh, Melbourne 3166, Australia

© Cambridge University Press 1989

First published 1989

Printed in Great Britain at the University Press, Cambridge

British Library cataloguing in publication data

Howse, Derek
Nevil Maskelyne, the seaman's astronomer.
1. Astronomy.
Biographies
I. Title
520'.92'4

Library of Congress cataloguing in publication data

Howse, Derek.
Nevil Maskelyne : the seaman's astronomer / Derek Howse ; with a foreword by Sir Francis Graham-Smith
p. cm.
Bibliography: p.
Includes index.
ISBN 0-521-36261-X
1. Maskelyne, Nevil, 1732–1811. 2. Astronomers–Great Britain–
–Biography. I. Title.
QB36.M378H69 1989
520'.92'4-dc19
[B]   88-15967   CIP

ISBN 0 521 36261 X

Contents

List of illustrations		vii
Foreword by Sir Francis Graham-Smith		ix
Preface		xi

PART 1 – PREPARATION
1	The Maskelynes of Purton	1
2	The longitude problem	9
3	The transit of Venus, 1761	18
4	Saint Helena, 1761	27
5	The Barbados trials, 1763–4	40
6	Astronomer Royal, 1765	53
7	The Royal Observatory and its instruments	62
8	The Harrison affair, 1764–7	74
9	The Nautical Almanac	85

PART 2 – ACHIEVEMENT
		99
10	Early years at Greenwich, 1765–9	99
11	The 1770s	113
12	Weighing the World – Schiehallion, 1774	129
13	The 1780s – and a new planet	142
14	The 1790s	163
15	The final years, 1800–11	184
16	Summing up	209

APPENDICES
A	The Maskelyne pedigree	212
B	Nevil Maskelyne's autobiographical notes	214
C	Maskelyne's expense accounts, St Helena, 1761–3	225
D	The 1765 inventory	229
E	Schiehallion instruments and equipment	232

F Memorial tablets in the Church of St Mary,
   Purton, Wiltshire   235

Glossary   237
Notes   242
Bibliography   260
Index   268

Illustrations

Fig.

	Nevil Maskelyne, by Russell, c.1776	Frontispiece
1.1	Ponds Farm, Purton Stoke	4
1.2	Margaret Clive, miniature, 1762	7
2.1	Chipping Barnet church	11
2.2	Sextant in use	16
3.1	Transit of Venus photo	19
3.2	Transit of Venus diagram	21
4.1	*Northumberland* East Indiaman	29
4.2	Transit of Venus engraving	35
4.3	Zenith sector diagram	36
7.1	Plan of Royal Observatory, 1788	64
7.2	Royal Observatory from south-east, c.1765	65
7.3	Old Royal observatory courtyard, 1982	67
7.4	Mural quadrant diagram	68
7.5	Transit Room, c.1785	69
7.6	Dollond telescope, c.1785	71
9.1	Nautical Almanac page headings	88–9
9.2	Lunar distance diagram	92
9.3	Lunar distance *pro forma*, 1772	95
10.1	Maskelyne's observing suit	101
10.2	Page of *Greenwich Observations*	103
10.3	Letter to Mason, 1769	111
11.1	Nevil Maskelyne by Downman, 1779	117
11.2	The Maskelyne crest and arms	128
12.1	Schiehallion from Loch Rannoch	133
12.2	The Attraction of Mountains diagram	134

12.3	Map of Schiehallion, 1774	135
12.4	The cairn at Braes of Foss	136
13.1	Nevil Maskelyne, by Van der Puyl, 1785	144
13.2	Sophia and Margaret Maskelyne, by Van der Puyl, 1786	145
13.3	The 1783 fireball, by the Sandbys	150
13.4	Royal Observatory, watercolour 1794	156
14.1	Margaret's watercolour of Royal Observatory, 1801	164
14.2	Margaret Maskelyne by Owen, c.1793	165
14.3	Joseph Banks with moon map, by Russell	166
14.4	The Coombe clock	168
15.1	Nevil & Sophia, miniatures by Byrne, 1801–3	194
15.2	Nevil, miniature by (?)Theed, 1801	194
15.3	Nevil, pastel by Russell, 1804	196
15.4	Sophia, pastel by Russell, 1804	197
15.5	Greenwich Park on Easter Monday, 1804	199
15.6	Letter to Vince, 1811	204–5
15.7	Purton church	206

Foreword

Professor Sir Francis Graham-Smith, FRS
ELEVENTH ASTRONOMER ROYAL

Nevil Maskelyne, fifth Astronomer Royal, established Greenwich as the source of navigational astronomy. His work led to the universal acceptance of the Greenwich zero of longitude. His achievements have had profound effects both on astronomy and in our national history. Astonishingly, however, there has been no biography of Maskelyne before Commander Howse undertook this work. We have been missing a feast, which is now supplied.

From the start of the Royal Observatory in 1675 to the publication of the first Nautical Almanac in 1767 is a span of 92 years. Maskelyne's predecessors had achieved much, but it is surprising that the observatory continued to be funded for so long without practical output. Maskelyne pulled the whole enterprise together, organizing his small resources to great effect. The result was the Almanac, together with the very practical Requisite Tables to facilitate its use. International collaboration was the key. The French astronomers and mathematicians provided the essential analysis of the Moon's motion, using the Greenwich observations which Maskelyne made available. The Almanac itself was adopted by many countries; the high esteem in which they held Maskelyne was reflected in his many foreign honours.

Maskelyne was Astronomer Royal for 46 years, and supervised the production of the first 49 issues of the Nautical Almanac. He presided over the famous tests of the Harrison timekeepers, he negotiated the publication of Bradley's observations, and he found the gravitational mass of the Earth from the Schiehallion experiment. Such a man deserves this fine biography from a dedicated historian such as Derek Howse, who knows navigation at first hand, and who knows the Royal Observatory and its history in every detail.

Preface

During the eighteenth century, astronomy's highest priority was in work connected with navigation, particularly 'finding the longitude'. Nevil Maskelyne was in the forefront of this work, very much at the 'sharp end', both in making and publishing the astronomical measurements the theorists needed, and in ensuring the mariner was not – in the words of King Charles II – 'deprived of any help the Heavens could supply, whereby Navigation could be made safer.'[1] To this end, he worked closely with astronomers and mathematicians all over Europe, particularly in France, despite the constraints imposed by war.

No biography of Maskelyne longer than a few pages has been written before and what has been written has generally concentrated upon scientific aspects. Luckily, a great deal of archival material has survived, both scientific and personal, particularly Maskelyne's own account books and pocket memorandum books where, for the last forty years of his life, he recorded every penny he received and spent, public and private, as well as recording daily happenings and notes, important and trivial, professional and personal, which tell us a great deal about how he lived and worked and of the people he met and corresponded with.

Though this book will be of particular interest to historians of navigation, astronomy and horology, it has been written as much for the general reader as for them. If I have sometimes included rather more detail – some may say trivia – than seems necessary, it is generally in an attempt to give some flavour of the way he and his friends lived. If there are things in this book which the general reader cannot understand, then I will have failed in my self-imposed task.

The book was conceived early in 1967, when I received a visitor in my office in the restored Meridian Building of the Old Royal Observatory in Greenwich Park, a few months before it was opened to the public by the then Astronomer Royal, Sir Richard Woolley. At that time, I was a curator in the Department of Navigation and Astronomy of the National Maritime Museum, whose Director and Trustees had become responsible for the observatory buildings – and for many of the instruments and clocks used there over three hundred years – when the astronomers departed for the clearer skies of east Sussex.

My visitor was Colonel Humphrey Quill, CBE, DSO, Royal Marines, Master of the Worshipful Company of Clockmakers that same year and author of the book *John Harrison, the man who found longitude*.[2] He brought with him some manuscripts written by Maskelyne, who lived in Flamsteed House for 46 years from 1765, making most of his important astronomical observations in the very building in which Colonel Quill and I were sitting. Among these papers were some 11 sheets in a folder inscribed by his daughter Margaret, 'Autobiographical notes in Dr Maskelyne's own writing'.

At the time, these appeared to be the notes for the autobiography that never got written (subsequent research reveals they were in fact probably prepared for an encyclopaedia entry). I knew no full-length biography of this person so important in the history of navigation and astronomy had been published, and therefore decided there and then that I must remedy that omission one day, believing that he would have approved of his own notes being used for such a project. In the event, it was not until 1983 that I was able to begin the serious research for that self-appointed task – after I had retired from my post at the museum, having worked for more than sixteen years in what had been Nevil Maskelyne's dwelling house and in his observatory buildings.

Acknowledgements

My first thanks must go to the Director and Trustees of the National Maritime Museum for granting me a Caird Research Fellowship after my retirement, specifically to produce this book. In the research phase alone, I visited more than a score of libraries and archives in Britain, France and the USA, and created more than eighty files of correspondence with individuals and institutions. None of this would have been possible for me without the generous financial support, special facilities, and great encouragement which stemmed from the fellowship.

Next, I must acknowledge the help given me by the late Professor Eric Forbes, who probably knew more about Maskelyne and eighteenth-century nautical astronomy than anyone else in the world. When he heard that I hoped to write this biography, he most generously passed on to me all the notes on Maskelyne and the Board of Longitude he had used in preparing his contribution to the *Dictionary of Scientific Biography* and his history of the Royal Observatory.[3] I have extensively and shamelessly used these

Preface

notes and his published work in writing this book and am very saddened that his untimely death in 1984 prevented him seeing the results of all his help.

In Chapter 11, I say that Maskelyne's family were caring people who had a proper sense of history. The present representatives of that family – Mr Nigel Arnold-Forster, his sister Mrs Vanda Morton, Mrs H. C. Arnold-Foster, and Mr Richard Masterman – have more than lived up to that tradition, giving me unstinting help and encouragement. I have also received welcome assistance from Canon R.H.D. Blake and Mrs Jane Ponting, both of Purton.

During the writing phase, I was helped enormously by those who accepted the chore of commenting upon the first draft chapter by chapter as they were written: Commander David Waters, Mr David Proctor, and Professor Stuart Malin, colleagues at the National Maritime Museum, past and present; and Miss Janet Dudley and Mr Adam Perkins of the Royal Greenwich Observatory. In addition, individual chapters were read by Dr David Cartwright and Mr Kenneth Harwood on the St. Helena story; Mr Stuart Leadstone and Professor J. F. Allen on the Schiehallion episode; Messrs Andrew King, Jonathan Betts, Charles Allix, and Richard Good on horology; and Dr George Wilkins on the Nautical Almanac.

My special thanks go also to Mr Alan Stimson, Miss Carole Stott, Dr Roger Morriss, and Mrs Elizabeth Wiggans of the National Maritime Museum; Mr Jacob Simon of the National Portrait Gallery; Mr Norman Robinson and his colleagues in the Royal Society library; Mr Peter Hingley of the Royal Astronomical Society; Dr Nicholas Rodger of the Public Record Office; Mr Andrew Cook of the India Office Library and Records; Mr J. d'Arcy of the Wiltshire Record Office; Mr Alan Kucia, archivist in Trinity College library, Cambridge; Mr Duncan Robertson, descendant of Maskelyne's helper on Schiehallion, and Mr Gregory Clarke; Professor Seymour Chapin in links with Lalande; and Mr John Bidwell of the William Andrews Clark Memorial Library of the University of California, Los Angeles, for introducing me to the joys of the microcomputer and word processing, which enabled me to index my notes on the six hundred or so Maskelyne letters I had found up to that time. And to the many, many other helpers whom space does not permit me to name, my most grateful thanks.

I am most grateful to Sir Francis Graham–Smith, eleventh Astronomer Royal, for doing me the honour of writing a foreword to this account of the life and work of his predecessor, the fifth Astronomer Royal. And finally, my thanks go to the fifth Astronomer Royal himself, for writing such good English, for being so systematic in his office work, and for writing such a legible, if not elegant, hand: these attributes, and his pleasant personality, made the writing of this book a most agreeable task for me.

Derek Howse
Sevenoaks, Kent

Part 1

PREPARATION

[These notes apply to the whole book, not just to Part 1.]

Notes

1. Technical terms mentioned for the first time are given in *italic* and defined in the Glossary.
2. The number following the name of a man-of-war, e.g. *Seahorse* 24, indicates the number of guns her hull was pierced for.
3. Birth and death dates are not generally given in the text but will be found in the index.

1

The Maskelynes of Purton

> Dr M. is the last male heir of an ancient family long settled at Purton in the County of Wilts, which from the name probably came from Normandy, where there is or was 50 years ago a family of that name Masqueline.[1]
>
> Maskelyne autobiography

So wrote Dr M. himself – the Reverend Nevil Maskelyne, FRS, DD, (1732–1811), fifth Astronomer Royal – and it seems fitting to start this account of his life with the very words with which he himself opened a brief unpublished autobiography which he wrote in 1800 when he was 68. This autobiography, all in his own hand, was handed down in the family for over 160 years before being deposited with his scientific papers at the Royal Greenwich Observatory in 1975. Elegantly written in beautiful English, it represents his own considered opinions on matters which are the main subject of this book, so one hopes he would approve of it appearing in print – complete in Appendix B.

Prologue

The village of Purton in Wiltshire is on low-lying ground north of the Marlborough Downs, some five miles west-north-west of the centre of today's sprawling town of Swindon, and more or less half way between Cricklade, four miles to the north, and Wootton Bassett, to the south.

The Maskelynes of Purton can be traced back at least eleven generations before the Astronomer Royal, to one William of Purton, whose son Robert Maskelynge was a grantee of lands from his father-in-law in 1435 in the reign of Henry VI. In 1679 at the age of 18, the Astronomer Royal's grandfather, Nevill Maskelyne (Nevil II), inherited considerable estates near Purton, and Lordship of the Manor of Cricklade, from *his* grandfather, Nevil I, who was named after his mother's ancestors, the

Nevilles of Abergavenny.² Since Nevil II's father William had died in London in 1676, Sir James Houblon (London merchant, brother to the first Governor and himself a Director of the Bank of England) had stood guardian to Nevil and his sister Jane, and this continued after Nevil inherited.

Nevil II soon settled down at West Marsh, Purton, as Squire Maskelyne, becoming Captain of a troop of horse (Militia) under Col Penruddock when he was 22, and Major of the Wiltshire Militia eight years later. The following year, 1692, at the age of 31, he married Anne Bath (or Bathe), daughter of the Vicar of Purton, by whom he had a large family. Writing in 1802, Margaret, Lady Clive, the Astronomer Royal's sister, described them thus:

I used to hear my Aunts relate that my grandfather's children were all dressed alike, changing the colour of their clothing, & that one particular year, as they walked [to church] in procession, an old woman sitting at the foot of Purton Church wall, cried out, in a treble voice, 'There go Squire Maskelyne's Yellow Hammers'. 5 sons and 5 daughters lived to grow up and sat at their parents' table. The old ladies used to take pleasure in this recollection.³

Their mother died in 1706 when the eldest of the ten children (Nevil again!) was only 14. Then, in 1711, their father died too, and the care of this large family fell upon the two eldest children, Nevil III, 19, and Jane, a year younger. Nevil III inherited the property but he had to provide for the education and maintenance of his brothers and sisters. In 1713, he became a clerk in the Secretary of State's office, first under the Tory, Henry St John, Viscount Bolingbroke, one-time MP for Wootton Bassett (presumably the reason for Nevil's appointment), then under his Whig successor, Charles, Viscount Townshend.⁴ However, Nevil III gave up that post in 1716 in favour of his 18-year-old brother Edmund. Two years later and still only 26, he was forced to sell the borough, hundred, and manor of Cricklade and much of the rest of his inheritance, including West Marsh, Purton. He retired to Down Farm, Purton and will appear later in our story.

At that time, his nine brothers and sisters – uncles and aunts of the Astronomer Royal – ranged in age from Jane, 25, to Wynn, 16. Of these, William, Elizabeth and Alice went to India, the girls soon finding husbands there. Of the others, Jane, Anne, and Sarah, all died unmarried. Of the two youngest, James Houblon Maskelyne became a lawyer at Clifford's Inn and will also reappear later in our story; and Wynn died without issue at the age of 36. (Appendix A shows the Maskelyne pedigree for the generations either side of the Astronomer Royal.)

Which leaves Edmund (1698–1744), the seventh child and third son of Anne and Nevil Maskelyne, and father of the Astronomer Royal. As we have seen, he followed his elder brother as one of seven clerks (the number varied) to the Secretary of State in 1716, and continued to hold this important post until his death in 1744. In today's terms, he could be thought of as the equivalent of a senior civil servant though he was actually a private employee of the Secretary of State, with no security of tenure, his superiors in the office being two Under Secretaries and a Chief Clerk. During his last twenty years, he served Thomas Pelham Holles, Duke of Newcastle, Secretary of State (South). He married Elizabeth, only child of John Booth of Chester, a distant cousin, some time before 1725. In 1735, he received the additional appointment as Deputy to one of the Clerks of the Signet, Charles Delafaye, who had been one of Newcastle's Under Secretaries until the previous year. This was an office concerned with the stamping of documents, by this date nearly a sinecure, which nevertheless brought in a salary and for which the Deputy would have received a percentage of all fees received.[5]

Nevil the astronomer is born

The four children of Edmund and Elizabeth Maskelyne were all born in a house in Kensington Gore, an area on the south side of Hyde Park, not far from today's Albert Hall – William (Billy) in 1725, Edmund (Mun to his future friends in India) in 1728, and Nevil IV, the future Astronomer Royal, (who seems to have had no nickname) on 5 October 1732 Old Style,[6] being christened at St. Martin-in-the-Fields, Westminster, on 27 October. Margaret (Peggy), the future Lady Clive, was born in 1735. The following year, the family moved to Tothill Street, Westminster, saving a walk of more than two miles each way to Edmund senior's office at Newcastle House in Lincoln's Inn Fields, to Whitehall, and to Westminster School where the two eldest boys had been elected King's Scholars (Newcastle was a Trustee of the school). In January 1741, Nevil followed his brothers to Westminster, becoming a Town Boy.

Then, in March 1744, their father died, leaving a somewhat meagre inheritance in trust for his three youngest children, the eldest, William, now a Cambridge undergraduate, being already provided for in the will of his maternal great-uncle. He took his BA in 1747 and became a Fellow of Trinity and unsuccessful candidate for the professorship of Hebrew in July 1753. He had inherited the estate of Purton Stoke from William Bathe the previous year, so, having taken Holy Orders (which he was

Fig. 1.1. Pond's Farm, Purton Stoke, Wiltshire. 'The Ponds', a sixteenth- and seventeenth-century farmhouse on an earlier moated site, the Nevil Maskelynes' 'country cottage', where the family used to spend six weeks or so every autumn. (Photo: Mrs Jane Ponting)

required to do as a Fellow of Trinity), he retired to the interesting moated house now known as Ponds Farm at Purton Stoke, later the Astronomer Royal's 'country cottage'. (Fig. 1.1.) Nevil later described him as a man of business but, in the words of one of Nevil's biographers, 'Relieved from all pressure of poverty, Wiliam made no name for himself in the world.'[7]

Through the patronage of the Duke of Newcastle, Elizabeth Maskelyne succeeded in obtaining for Edmund, her second son, the post of writer in the East India company.[8] In 1744, at the age of seventeen, he sailed for India, arriving at Madras a month or so before his future brother-in-law, Robert Clive, who was nearly three years older than Edmund.

Nevil's school-days

In the meantime, Nevil was still at Westminster. Continuing the quotation at the beginning of this chapter, he was to write this in 1800:

He . . . received his classical education at Westminster School, & instructions in writing & arithmetic during intervals of school from other masters, which proved of the greatest service to him in his future scientific pursuits. So much time is taken up in attaining the learned languages themselves that it cannot be expected that a great stock of knowledge can be acquired at school for forming the mind; but this he supplied by reading with avidity our best english authors. From occasional discourses in the family, he became eager to see the effects of telescopes on the heavenly bodies, and to know more of the system of the universe. Observing the great eclipse of the sun in 1748 with the late Mr Ayscough in an unusual manner by means of the sun's image projected through a telescope on a white screen in Camera Obscura[9] added fresh spur to his astronomical desires, and from this time he returned himself seriously & closely to the study of the two kindred sciences of Optics & Astronomy, which he has pursued with unwearied diligence ever since.[10]

He seems to have been rather a swot and perhaps a bit of a prig! Delambre, in his *Eloge* to Maskelyne written in 1813, amplifies this:

. . . but what attached him to the prosecution of these studies was the eclipse of the sun of 1748, of which ten digits were eclipsed at London. It is very remarkable that this eclipse produced the same effect on the mind of Lalande [the future French astronomer], who was only three months older than Maskelyne; and it may with truth be observed that no celestial phaenomenon was ever more useful to science than this eclipse, which gave her two such very distinguished astronomers, who pursued this science under different views, each taking the department most agreeable to his own taste. One wrote largely in all branches of astronomy, and instructed others with great success, but made very few observations; the other has written comparatively little, but his numerous observations are universally acknowledged to possess an unrivalled degree of accuracy.[11]

Later, Nevil Maskelyne in Greenwich and Joseph-Jerôme-Lefrançais de Lalande in Paris became great friends and cooperated in the production of their respective ephemerides, despite the difficulties brought on by war. The Mr Ayscough mentioned by Maskelyne cannot be identified for certain. He could have been the optician and instrument maker James Ayscough at the Sign of the Golden Spectacles in Ludgate Street, or possibly the Rev. Samuel Ayscough of Bedford Row, London, whose address he noted in 1786.[12] Certainly, their observations were never published.

In the winter of 1748–9, Elizabeth died – 'Poor Neice Maskelyne died of a Palsey' is a note in the diary of her aunt, Mrs Katherine Howard[13] – and Nevil and his brothers and sister became orphans in respect of both parents, as had the previous generation nearly forty years before. William and Edmund were already settled. Nevil became a boarder in Vincent Bourne's house in Great College Street. At Westminster School with him were many who were to become as famous as himself – Rockingham, afterwards Prime Minister; William Hamilton, husband of Nelson's Emma; William Cowper, the poet; Charles Churchill, a lesser poet;

Warren Hastings, of India fame; and Edward Gibbon, the historian. The Maskelyne arms are still to be seen, painted on the panels of the Great Schoolroom, where the entire school was taught in the days of the Maskelyne brothers.[14]

To continue Nevil's own account of his school days:

> Many mathematicians have become Astronomers from the facility Mathematics gave them in the attainment of Astronomy; but here the love of Astronomy was the motive of application to mathematics without which our Astronomer soon found he could not make the progress he wished in his favorite science; in a few months, without any assistance he made himself master of the elements of Geometry & Algebra. With these helps he soon read the principal books in Astronomy & Optics & also in other parts of natural philosophy, Mechanics, Pneumatics & Hydrostatics. The considerable progress he had made in these sciences led him naturally to the University of Cambridge, where these studies were then as they are still, peculiarly encouraged.[15]

Margaret Maskelyne

Meanwhile, Margaret, orphaned in 1748 at the age of 13, had gone to live with her aunts in Wiltshire, attending Mrs Saintsbury's school in Cirencester.[16] She kept up a correspondence with brother Edmund in India, who, as a matchmaker, tried to persuade her to come and join him there. Indeed, he had it all arranged. In February 1751, their cousin Eliza Fowke (née Walsh) wrote from Madras to her aunts in England, telling them that her cousin, Mun Maskelyne, had 'laid out a husband for Peggy if she chooses to take so long a voyage for one, that I approve of extremely, but then she must make haste, as he is in such a marrying mood that I believe the first comer will carry him.'[17] (Later that year, she was so impressed with a letter Margaret has written her in French that she offered to pay for her to go to school in London to perfect her French and her dancing.)[18] The potential husband referred to was a colleague of Edmund's called Robert Clive. In March 1752, Edmund renewed his pleas:

> Dear Peggy,
>
> I was favoured with yours of the 20th August 1751, by the Dorrington, on the 14th instant, and hope your declining my proposal for tripping it this way proceeds from some more agreeable views at Home, as otherwise I can't blame you for it: matches in this Country generally proving so vastly superior to what are made in Europe.
>
> I am extremely glad you left our Uncle Nevil in good health; whom I do

myself the Pleasure of writing to, by this conveyance; and hope Mrs. Fowke who quitted us in October last, is so too.

As I have given you my advice supported by the most Solid Reasons I could urge; I leave it entirely to you to determine for yourself, only begging your assurance that whenever we have the happiness of meeting

You will find
a sincerely affectionate brother
in
Edmd. Maskelyne

Dericota
March 20th 1752

P.S. You leave me quite in the Dark as to where the young Astrologer Nevil is – [He was by now at Trinity College][19]

Though she must have already sailed for India before this letter reached England, Edmund's earlier persuasive powers worked and his predictions proved right. Margaret, not 17 until the following 6 November, sailed for Madras with eleven other young women, landing in Madras in late June 1752. She stayed with her brother in his house in Veperi, west of the Fort; he had been captured by Indian troops allied to the French in July 1751 but was now living in Madras on parole.[20]

Robert Clive, fresh from his triumphs at Arcot and Trichinopoly, who had seen her miniature portrait in Mun's rooms, fell for this slender girl with dark hair and large eyes, despite being 'prevented from being beautiful by her too large nose and too thick eyebrows'[21] (Fig. 1.2.) Robert and

Fig. 1.2. *Margaret Clive*, 1762, aged 27. From a miniature pencil sketch inscribed *W. S. King/Lady Clive/Esq:/May 25 1762*, in possession of Nigel Arnold-Forster, Esq. On that date, the Clives were in England. (Photo: author)

Margaret were married in Madras on 18 February 1753, when he was twenty-eight and she seventeen. The Clives returned to England later the same year, arriving in October and setting up house in Queen Square, Ormond Street, thus providing a London *pied à terre* for Astrologer Nevil, now in his final years as an undergraduate at Cambridge.

In April 1756, the Clives sailed once more for India, taking with them Margaret's 16-year-old cousin Jenny Kelsall, who they successfully married off to Thomas Latham, captain of one of the naval squadron in Indian waters. By July 1760, when they next set foot in England, Clive was not only famous for his military accomplishments in Bengal, but was also very rich. We shall, of course, meet them later in the story.

2

The longitude problem

Nevil at Cambridge

Leaving Westminster school in July 1749, Nevil Maskelyne entered Catharine Hall, Cambridge, on 5 November[1] as a sizar, a scholar who paid reduced fees but who had to perform certain menial tasks in return. (They used to have what was left at the Fellows' table, because it was their duty at one time to wait on the Fellows at dinner, when each Fellow had his own sizar.) In July 1750, he migrated to Pembroke College where he matriculated. At the age of 20, he moved again, to his brother William's college, Trinity, where he became a pensioner on 13 December 1752, a scholar in 1753, and took the Mathematical Tripos for the degree of Bachelor of Arts in 1754, graduating as seventh wrangler. (The first of the junior optimës in the same examination was Erasmus Darwin, grandfather of Charles Darwin.)

While Nevil was up at Cambridge, only two documents have been found which throw any light on his private affairs: a common-place book dated at one end Pembroke College, December 29 1750, in which he has written a number of poems and versions of psalms, one dedicated to Mr Glasse from Catharine Hall, while at the other end of the book are written notes on mathematics, optics, etc., started at Cambridge but continued at Greenwich;[2] and an attendance list for the Westminster Club Supper at the Three Tuns Tavern on 28 November 1753 with William Maskelyne as the President and young Nevil as one of the diners.[3]

One of the poems in the common-place book has a rather modern theme and throws some light on his character at that time:

> UPON THE AUTHORESS OF THE LADY'S MAGAZINE
> Too long had Man usurp'd imperious sway
>   O'er the Fair Sex and made the weak obey:
> Yet knew that power by force not right obtain'd
> Should be by Art & Subtlety maintain'd
> Hence learning to their Schools alone confin'd
> T'improve their reason & inlarge their mind

> While poor weak women learn to spend their days
> In Balls in Cards in Novels & in Plays.
>
> Amelia Carolina seeks renown
> Her Sex's right asserting & her own
> To either Sex unfolds the shining Page
> At once to please & to instruct the Age
> To latest times she shall extend her fame,
> Who's mind's harmonious as her liquid name
> Graia's Tenth Muse Sappho shall yield to you
> For sprightly wit & sense & virtue too.[4]

Did Nevil write this with tongue in cheek? Research has so far not revealed the identity of the blue-stocking Amelia Carolina.

Maskelyne was anxious to become an astronomer – and an important one. The first real hurdle was the Mathematical Tripos which he had just obtained. Next, it would be an enormous advantage to be elected a Fellow of his college: not only would this be a mark of academic distinction but it would also pay a small stipend and give him free board and lodging in College until he married. To be elected, a candidate had to pass a searching oral examination in mathematics, classics, philosophy and theology by the Master of the College and eight senior Fellows: once elected, a Fellow had to take holy orders within a short time.

In 1755, awaiting the appropriate moment to apply for Fellowship, he decided to anticipate the requirement for taking holy orders. The church at Chipping Barnet in Hertfordshire (Fig.2.1) some ten miles north of London, was served by a curate appointed by the rector of East Barnet, then a one-time Fellow of Trinity called Samuel Grove. At Michaelmas 1755, Maskelyne was ordained to the curacy of Chipping Barnet.[5]

In fact, as the senior Westminster scholar in 1756, he was virtually assured of his election to the Fellowship as there was a clear link between Westminster and Trinity, although this was not mentioned in the College statutes; at this time, almost half of the Fellows of Trinity had been educated at Westminster. So, on 2 October 1756, Maskelyne was duly elected a Fellow of Trinity. He was elected to his Major Fellowship on 6 July 1757 after he had proceeded to the M.A. degree.[6]

Discovering the longitude

It was about this time that Maskelyne obtained an introduction to the Astronomer Royal, James Bradley, whose reputation stood high among world scientists, particularly for his discovery of the *aberration of light* (See

The longitude problem

Fig. 2.1. *Chipping Barnet Church*, Hertfordshire, where Nevil Maskelyne was curate from 1755 to 1763. From a watercolour attributed to Thomas Rowlandson in Hertfordshire County Record Office. (Photo: HRO)

Glossary) before going to Greenwich, and for his discovery of the *nutation of the Earth's axis* soon after he became Astronomer Royal in 1742. Since that date, however, his chief work was concerned with the direct use of Astronomy in the service of Navigation, fulfilling the terms of his royal appointment, '. . . forthwith to apply yourself with the most exact Care and Diligence to the rectifying the Tables of the Motions of the Heavens, and the Places of the fixed Stars, in order to find out the so much desired Longitude at Sea, for the perfecting the art of Navigation.'[7] This same directive had been given to John Flamsteed, the first Astronomer Royal, by King Charles II in 1675, and was repeated for Edmond Halley, the second in that post, by George I in 1720, and for Bradley by George II in 1742.

This whole subject of finding longitude at sea had been polarized by the British Government's Longitude Act of 1714, offering huge rewards up to £20 000 'for a due and sufficient Encouragement to any such Person or Persons as shall Discover a proper Method of Finding the said Longitude', payable regardless of nationality 'as soon as such Method for the Discovery of the said Longitude shall have been Tried and found Practicable and useful at Sea. . .'.[8] The Act appointed Commissioners, familiarly known as the Board of Longitude, to administer its provisions.

Headed by the First Lord of the Admiralty, they included the Speaker of the House of Commons, various admirals and Admiralty officials, mathematics and astronomy professors from Oxford and Cambridge, the President of the Royal Society, the Master of Trinity House, and, of course, the Astronomer Royal.

Much of Maskelyne's future career was bound up with the provisions of that Act, so it is worth a brief diversion to consider its more important provisions:

(a) *The Commissioners.* Originally 23 were appointed, though seldom more than seven or eight came to meetings. For major awards, a majority vote of these was needed.

(b) *Awards for experiments.* If a proposal had promise, awards up to £2 000 could be made without further ado by a quorum of five Commissioners.

(c) *Major rewards, accuracy and amounts.* The act went on to state the awards that could be given:

> And for a due and sufficient Encouragement to any such Person or Persons as shall Discover a proper Method of Finding the said Longitude, be it enacted by the Authority aforesaid, That the First Author or Authors, Discoverer or Discoverers of any such Method . . . shall be entitled to, and have . . . a Reward, or sum of Ten thousand Pounds, if it Determines the said Longitude to One Degree of a Great Circle, or Sixty Geographical Miles; to Fifteen thousand Pounds, if it Determines the same to Two Thirds of that Distance, and to Twenty thousand Pounds, if it Determines the same to One half of the same Distance. . .

> Points to note are: (i) that the major reward can only be awarded once; (ii) no two economic historians agree about the value in today's money of £20 000 in 1714, but almost all put it above a million pounds, while some put it as high as £3.5m;[9] (iii) as for the accuracy needed, one must beware of thinking the requirement was finding longitude to one degree of longitude: the unit stipulated was a degree of a great circle, or 60 geographical miles, which, for our purposes, can be considered as a degree of latitude, or 60 nautical miles: because of the convergence of the meridians between equator and poles, the length of a degree of longitude gets smaller with increasing latitude: the accuracy of longitude needed for the highest award, 30 geographical miles (or 30 minutes of longitude on the

The longitude problem 13

equator), is equivalent to a difference of longitude of about 0°32' in the West Indies, about 0°47' in the latitude of the English Channel.

(d) *Trials*. Half the reward would be paid when the Commissioners agreed that:

> any such Method extends to the Security of Ships within Eighty Geographical Miles from the Shores, which are Places of the greatest Danger.

The other half would be paid (provided the requirement in (e) below was satisfied):

> when a Ship . . . shall actually Sail over the Ocean, from Great Britain to any such Port in the West-Indies, as those Commissioners . . . shall Choose or Nominate for the Experiment, without Losing their Longitude beyond the Limits before mentioned.

(e) *Requirement*. The reward could then be paid:

> as soon as such method for the Discovery of the said Longitude shall have been Tried and found Practicable and Useful at Sea, within any of the Degrees aforesaid.

Many future disputes revolved around the interpretation of the words 'practicable' and 'useful'.

(f) *Minor awards* could be made at the Commissioners' discretion for proposals which, though not qualifying for a major reward, nevertheless were found 'of considerable Use to the Publick.[10]

By 1755, it was generally agreed that the solution lay with astronomy – as opposed to, say, geomagnetism – possibly aided by the technology of clockmaking. The theory of the astronomical solution had been known since early in the sixteenth century. Local time and longitude are so related that a difference in one is always equal to a difference in the other; therefore, to find the difference of longitude between one's own meridian and that of some reference meridian – shall we call that Greenwich? – all one has to do is to find the difference between the local time of the ship and the local time on the reference meridian *at that very same moment* – which we will call Greenwich time – this difference in hours and minutes of time being generally converted into degrees and minutes of arc. The practical solution was, however, somewhat more difficult.[11]

The invention in 1732 of Hadley's reflecting quadrant (sometimes called an octant, because it had a 45° arc measuring to 90° by double reflection) made it comparatively easy for the observer to find his local time by measuring the altitude of a heavenly body some distance from the meridian, and then solving a simple spherical triangle. But how could he know what time it was at Greenwich at that same moment? In the eighteenth century, two possible solutions were mooted: carry Greenwich time with you in the form of one or more timekeepers – this came to be called *the chronometer method* – or use the Moon as a sort of clock, measuring her place against the background of the stars (she moves comparatively fast, approximately her own diameter in an hour), and then looking up in an almanac what Greenwich time it was predicted she would be in that place – which came to be called *the lunar-distance method*. In the mid-1750s, however, neither method was as yet practicable: a marine timekeeper with the requisite precision and reliability had not been made; and the position of the Moon against the background of the stars could not be predicted to the required precision. But both were to be achieved within fifteen years, thanks largely to the incentives promised by the Longitude Act, just as Parliament had intended. And two people intimately connected with these developments were Bradley and Maskelyne.

When Maskelyne first met Bradley, the Board of Longitude was considering proposals concerning each of these methods. To perfect his marine timekeepers, John Harrison had already received from the Board a total of £2 250 between 1737 and 1755; in 1757, he received a further advance of £500 to complete the adjustment of his third large timekeeper and a new large watch which promised well. In that same year, the Göttingen astronomer Tobias Mayer sent to Admiral Lord Anson, First Lord of the Admiralty and chairman of the Board of Longitude, a set of lunar tables which he claimed could be used to predict the Moon's place for the lunar-distance method. He also submitted a portable repeating circular optical instrument for angle-measurement. The tables were referred to Bradley who compared the predictions with actual observations taken at Greenwich and reported that there were no differences greater than 1.5 arc-minutes, which should allow longitude determination to about half a degree. Sea trials of these tables and of the reflecting circle were carried out by Captain John Campbell, R.N., who had been the then Commodore Anson's master in the *Centurion* during the circumnavigation of 1741, when, because longitude could not be measured, Anson failed to find the island of Juan Fernandez when he expected, causing the ship to run out of fresh provisions, resulting in many lives being lost through scurvy.

Campbell's first trials were in 1757 when he was commanding the *Essex*

70 (wrecked in action in Quiberon Bay), then in the *Royal George* 100 in 1758–9, when he was Admiral Hawke's Flag Captain. He had taken a series of eleven observations in June, July and September 1759 within sight of Ushant – during Hawke's blockade of Brest preceding the Battle of Quiberon Bay – and these showed a spread of less than a degree of longitude.[12] These trials were inconclusive because of the constraints imposed by the Seven Years War. Campbell thought Mayer's repeating circle too heavy and unwieldy so, with the cooperation of the instrument maker John Bird, he developed the brass sextant we know today (Fig.2.2). With an octant measuring angles up to 90°, it was only possible to observe the distance between the Moon and the Sun on eight days each lunation; with a sextant measuring to 120°, the same observation was possible, weather permitting, on fifteen or sixteen days each lunation.[13] With a circular instrument, angles greater than 120° can be measured, and Mayer claimed this meant more observing days each lunation, but this advantage was to some extent nullified in practice because of the difficulty of measuring angles larger than 120° at sea.

Nevil the curate

Exactly how Bradley and Maskelyne met, we do not know, nor have we any details of how they cooperated, though it was reported that Maskelyne helped the Astronomer Royal prepare his tables of atmospheric refraction. Maskelyne took his degree of Master of Arts in 1757, and then, on 16 January 1758, the following certificate was presented to the Secretary of the Royal Society:

The Revd. Nevil Maskelyne A.M. Fellow of Trinity College in Cambridge a person well versed in Mathematical Learning and Natural Philosophy being desirous to become a Fellow of this Society is recommended by Us of our own personal knowledge as well qualified and likely to prove a worthy and usefull Member. 16 Jan.1758

> Rd. Hassell
> Ja: Bradley
> Tho. Simpson
> J Robertson
> Wm. Mountaine
> Francis Blake

This certificate lay on the table at ten consecutive Society meetings and then, on 27 April 1758, he was balloted for and elected Fellow of the

Fig. 2.2. Taking The Altitude of a heavenly body with a marine sextant. From the log of the *Owen Glendower*, 1846-7. (NMM MS Log M1.) (Photo: NMM)

Royal Society, permitting him to put the coveted initials F.R.S. after his name. Of the sponsors, Richard Hassell, a graduate of Queen's College, Cambridge, lived at Chipping Barnet and it was in his house there that Maskelyne was then lodging; James Bradley, whom we have already met; Thomas Simpson was Professor of Mathematics at the Royal Military Academy at Woolwich; John Robertson, mathematician, was mathematical master at the Royal Naval Academy at Portsmouth who had published *The Elements of Navigation* in 1754 (it was to go through seven editions in fifty years); William Mountaine was a writer and teacher of navigation; and Sir Francis Blake, Bart., was a mathematician.

As to Maskelyne's activities in Chipping Barnet, either pastoral or scientific, we know very little. He officiated at marriages between January 1756 and March 1758 and again in April 1763. On 17 June 1758, he set up two thermometers and a barometer in Richard Hassell's hall and garden, being visited by Dr John Bevis and Mr Bennet with Captain Ashurst's portable barometer, which they then took back to Bevis's house in London; from then until 17 October he kept a meteorological journal.[14] Several surviving letters are dated from Gray's Inn, London, and in April 1763 he was still staying with Hassell.

In the June 1758 number of *The Gentleman's Magazine*, a certain 'N.M.' wrote to Mr Urban saying that he had seen a comet on 20 and 23 June with his reflecting telescope, giving its approximate positions in the sky; 'This appearance I took the earliest opportunity of intimating to those gentlemen who hold the first rank in the astronomical world, either for science or observation.'[15] Replying, 'B.I.' said that this was most unlikely to be the comet of 1682 – Halley's comet – the return of which was imminent. It is not impossible that 'N.M.' was our subject and that 'B.I.' was John Bevis, or possibly James Bradley.

But it was at a meeting of the Royal Society on 5 June 1760 that Maskelyne began to see his way ahead in an astronomical career.

3

The transit of Venus, 1761

> The expected transit of the planet Venus over the sun in 1761, a phaenomenon which does not happen above once or twice a century, occupied the attention of Astronomers during the preceding year to prepare for the due observation of it, to which they were stimulated by the exhortations of the late Dr. Halley published in the philosophical transactions, who had observed the exit of Mercury from the sun at the Island of St. Helena in 1672 apparently instantaneous & thence concluded that the like phaenomenon of Venus might be observed the same, from which he concluded that if both the entrance & exit of Venus were observed by two observers in proper distant stations on the globe, that great *desideratum* in Astronomy, the sun's parallax, might be determined to the 500th part of the whole. Altho' Dr. Halley was mistaken in some particulars, partly owing to the imperfect Astronomy of the planet at that time, so that the transit of Venus could not be expected to determine the parallax as near[?] as he had supposed of so great advantage, yet still it afforded the best means of determining it that had yet been put into practice.
>
> Maskelyne autobiography

Twice every 113 years, the planet Venus passes directly between the Earth and the Sun. For a few hours, it appears silhouetted against the bright solar disk (Fig.3.1). The transits occur in pairs, separated by eight years. The first time this was observed was by the North Country astronomers Jeremiah Horrocks (who had predicted the transit) and William Crabtree on 24 November 1639.

In 1677, Edmond Halley, Bradley's predecessor as Astronomer Royal, had gone to St Helena to catalogue those southern stars which could never be seen from England, in order to complement the cataloguing of the northern skies being undertaken by John Flamsteed, first Astronomer Royal, at the newly-founded Royal Observatory at Greenwich. While Halley was there, he observed a transit of Mercury (which happens much more frequently than the transit of Venus) and realised that measure-

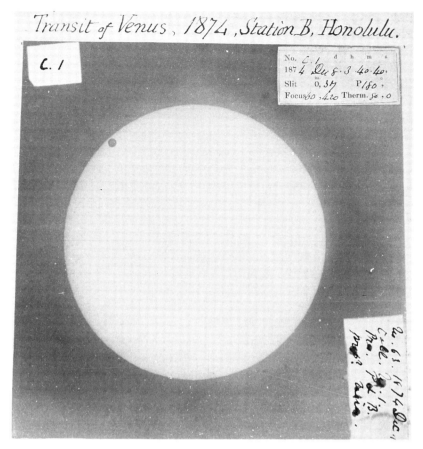

Fig. 3.1. The Transit of Venus across the Sun's disk, 1874. From a photo taken by astronomers from the Royal Observatory, Greenwich. (Photo: NMM)

ments taken during such a transit could in principle provide a method of finding one of the most fundamental quantities in astronomy, the distance of the Earth from the Sun – known today as the *Astronomical Unit* – from which the distance-scale of the whole solar system could be derived.

In 1691 and again in 1716 he published his ideas in the Royal Society's *Philosophical Transactions*,[1] pointing out that a transit of Venus, rather than that of Mercury, could provide data to the required precision, predicting that there would be another transit on 25 May 1761, and suggesting places without existing observatories where it would be most valuable to send expeditions to observe from – Northern Norway, India, Burma, the East Indies, and Hudson's Bay.

The quantity actually measured, from which the Astronomical Unit can be derived, is *solar parallax*, defined as the angle subtended by the Earth's equatorial radius at the mean distance of the Sun, or, more simply, the radius of the Earth as seen from the Sun. If we say that the Sun's parallax is 10 arc-seconds, it is the equivalent to saying that, seen from the Sun, the Earth has an apparent radius of 10 arc-seconds. The theory behind the transit observatione can be seen in Fig.3.2. An observer at A in the northern hemisphere (say Greenwich) would see Venus crossing the sun's disk on track aa'. An observer at B in the southern hemisphere (say St Helena) would see Venus crossing at bb'. Halley pointed out that solar parallax could be obtained by finding the difference between the lengths of the two chords aa' and bb', and that the lengths of the chords could be easily obtained by timing the moments of ingress and egress of the planet on the Sun's disk at each place. The greater the difference in latitude between A and B, the greater the difference between the lengths of the chords, and consequently the greater the precision of the result.

Though Halley knew that he himself would not live to see the next pair of transits (in 1761 and 1769), he urged that every advantage be taken of these opportunities, and that observations should be planned for as many places as possible, to mitigate against the effects of bad weather at any one site. 'Therefore,' he wrote, 'I strongly urge diligent searchers of the Heavens (for whom, when I shall have ended my days, these sights are being kept in store) to bear in mind this injunction of mine and to apply themselves actively and with all their might to making the necessary observations. And I wish them luck and pray above all that they are not robbed of the hoped-for spectacle by the untimely gloom of a cloudy sky; but that at last they may gain undying glory and fame by confining the dimensions of the celestial orbits within narrower limits.'[2]

Halley's pleas did not fall upon deaf ears, though most of those ears were in France rather than Britain in the years before 1761. Particularly influenced by Halley was Joseph-Nicolas Delisle, who had been appointed *astronome de la Marine* in 1747 on his return from a 22-year stay in St Petersburg. He had visited Halley at Greenwich in 1724 and probably discussed the matter. One of the difficulties with Halley's method of observation was that it required the observer to time very precisely the moments of contact of the planet with the Sun's disk at both ingress and egress, but at any one transit, the parts of the Earth's surface where both ingress and egress could be seen were very much restricted; in many areas, the ingress took place before sunrise (so only the egress could be observed), while in others, the egress took place after sunset (so only the ingress could be observed). Delisle devised a method whereby it was

The transit of Venus, 1761

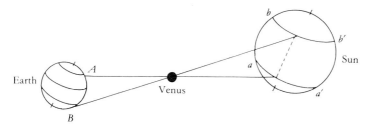

Fig. 3.2. The Transit of Venus. Observers on Earth at *A* and *B* see the silhouette of Venus crossing the Sun's disc along tracks *aa'* and *bb'* respectively.

possible to obtain similar results by combining the observations of two stations in similar latitudes, one of which could see only the ingress, the other only the egress, *providing* the difference in longitude between the two stations could be accurately found, a serious additional requirement except at established observatories.[3]

In England, there was little enthusiasm initially, though the *Gentleman's Magazine*, for example, did publish letters and articles in 1758 and 1759,[4] while from May 1760, the East India Company sent to all its Presidencies 'at the Sollicitation of some members of the University of Cambridge', instructions by the Rev. John Michell, Fellow of Queen's College, on how the transit should be observed.[5] (In view of subsequent events, one wonders whether Maskelyne, who knew Michell well, had a hand in this.) However, perhaps because of the Seven Years' War with France (which began in 1756), no one in authority in Britain began to make any preparations for expeditions such as Halley had suggested until just a year before the next transit was due (the predicted date was now 6 June 1761), barely enough time in view of the travelling time needed. The first official thought in England seems to have been given on 5 June 1760 when a printed memoir dated 27 April 1760 from Delisle to the French King was read to the Royal Society, having been brought to London from Paris – and considerably delayed by wartime hazards – by the Jesuit Father Roger Boscovich, Professor of Astronomy at the Roman College.[6] On a map of the world, Delisle had plotted those areas where the transit would be visible, showing the times where it would begin and end. He suggested places for observation additional to those suggested by Halley. Meanwhile, in France, considerable preparations were being made, one observer, Guillaume-Joseph-Hyacinthe-Jean-Baptiste Le Gentil de la Galaisière, having already sailed from Brest on 26 March for Pondicherry, while plans were in hand for expeditions to Siberia and the Cape of Good Hope or the south Indian Ocean.[7]

At a Royal Society meeting a fortnight later, on 19 June, 'a motion was made by the Rev. [Charles] Moss, seconded by Mr. [Daniel] Wray, and carried *nemine contradicente*, that it be recommended to the Council to think of, treat with, and at their discretion to conclude agreements with proper persons to go to proper places, in order to observe the expected Transit of Venus over the Sun: and that Dr. Bradley be particularly requested to favour such persons so to be deputed by the Council, with his advice and instructions for their conduct upon the occasion.'[8] The Council duly met to begin their preparations on 26 June, under the President, the Earl of Macclesfield, a distinguished astronomer in his own right. As stations for British observers outside Britain, the Council chose St Helena (where only egress would be visible), and, if there was time for the observers to get there, Bencoolen in Sumatra (where both ingress and egress would be visible). Both of these places were at that time under the control of the British East India Company though, by the date of the transit, both Bencoolen and Pondicherry had changed hands, the former being captured by the French, and the latter by the British. Bradley, the Astronomer Royal, was asked to investigate whether the necessary instruments could be hired and James Burrow (a later President) was asked to enquire from the East India Company what assistance they could give in passages and accommodation. Council members were asked to consider who should be the observers.

At the main Society meeting the following week, on 26 June, Maskelyne read a paper entitled 'A Proposal for Discovering the Annual Parallax of Sirius'[9] In this, he suggested that whoever went to St Helena for the transit of Venus – could he have thought this might be himself? – should also make observations of the star Sirius over a period of a year, to find *its* parallax, and thence its distance, using as a base line the diameter of the Earth's orbit, which the Earth takes six months to cross. (This is a different kind of parallax from the solar parallax described above, which has the Earth's radius as a base line.) Sirius is the brightest star in the sky and therefore probably one of the closest, with a correspondingly large and hence more easily measurable parallax. Furthermore, because Sirius passes almost overhead at St Helena, far more precise observations were possible there than in England, due to the fact that the effects of atmospheric refraction (which vary not only with altitude but also with air temperature and pressure) reduce to zero when looking right overhead at the zenith. And only with very precise measurements indeed could any results be expected. Maskelyne pointed out that, if successful, such an experiment would give incontrovertible proof of the Copernican system, still not firmly proved two hundred years after Copernicus himself had first proposed it. (When Bradley discovered aberration, he was trying without success to find the parallax of the star Gamma Draconis which passes nearly overhead in England.)

Maskelyne's paper was followed by one entitled 'On the approaching transit of Venus'[10], by Boscovich, just proposed for Fellowship of the Society, sponsored by, among others, Macclesfield, Bradley, and Maskelyne.

At the Council's next meeting on 3 July, Burrow reported that, while the East India Company had promised to do everything possible to help, there was no way they could get observers to Bencoolen in time in any of their ships, though the Dutch might be able to help. Then Bradley said that, as there were no suitable instruments for hire, he was presenting an estimate of the cost of the instruments needed, drawn up by Maskelyne – who obviously reckoned the job of observer was now his – two 2ft reflecting telescopes for St Helena, one for Greenwich; one each Dollond micrometers for St Helena and Greenwich; a clock with a compound pendulum and an 18in quadrant for rectifying the clock for St Helena; a total of £185; and, in addition, a 10ft radius zenith sector for the Sirius observations, at £100. The Council then considered the estimated costs for instruments and of observers' travel and keep for St Helena. This came to £685:

... which sum being judged greater than was convenient for the Society to expend, a memorial was drawn up, to be presented to the Lords of the Treasury in the following terms:

The MEMORIAL of the President, Council, and Fellows of the Royal Society of London for Improving Natural Knowledge: HUMBLY SHEWETH,

THAT WHEREAS the French, and Several other Courts of Europe, are now sending proper persons to proper places in various parts of the world, to observe, for the Improvement of Astronomy, the Transit of Venus over the sun, which will happen on the Sixth of June next;

AND WHEREAS this Nation is more immediately concerned in this Event, predicted in the last century by an Englishman, Dr. Halley, his Majesty's late Astronomer Royal, and never observed but once before since the World began, and then only by another Englishman, the ingenious Mr. Horrox;

AND WHEREAS the Expences of this most laudable undertaking, in which the Honour of this Nation is thus principally concerned, appears, upon an Estimate of the Charges thereof, to be Eight Hundred Pounds, if only two persons be sent with the necessary instruments to the island of Saint Helena only; and if the like number be also sent to Bencoolen, which is very much to be desired, will amount in the whole to near double that sum: the least of which sums is disproportionate to the circumstances of this Society:

The said President, Council, and Fellows, do therefore humbly request your Lordships to intercede with his Majesty, that he would be most graciously pleased to enable them to carry the said design into Execution, in such manner as to his Majesty in his great Wisdom shall seem proper.[11]

This splendid appeal to patriotism so soon after the *annus mirabilis* of 1759 succeeded admirably and the Duke of Newcastle, by now Prime Minister, informed the President on 10 July that he had laid the Memorial before King George II (he was to die later that year) who had ordered a warrant to be prepared for a grant of £800 for the St Helena expedition. A little later, another warrant for £800 was issued for Bencoolen. Though fees at the Treasury were remitted, fees at the Exchequer meant that the Society lost some £40 on each transaction.

Told on 14 July about the King's response to the Memorial, the Council appointed Maskelyne to lead the St Helena expedition with a gratuity of £150 plus expenses, with Charles Mason, Bradley's assistant at Greenwich, as second observer at £100. They also told Maskelyne to order the necessary instruments. Bradley promised to lend an 18-inch quadrant. Only five weeks after Delisle's paper was read, one expedition was fully organized, a second well on the way. The Royal Society had done well. But what a pity preparations were not begun a year earlier, in spite of the war.

On the day the Memorial was prepared, Maskelyne dined as a visitor with the Royal Society Club, a sure sign that he had joined the scientific establishment of the day. Sometimes known as the Club of the Royal Philosophers but at this time generally referred to as the Mitre club, this exclusive body, founded about 1731 by Edmond Halley and formalized in 1743 after his death, met weekly for dinner at the Mitre Tavern in Fleet Street before the Thursday meetings of the Royal Society in nearby Crane Court near Fetter Lane.[12] There were generally about twenty diners, including visitors, and many important decisions affecting the Royal Society's policy were taken at these dinners. The week after Nevil's first dinner, his elder brother William was a fellow visitor. Before leaving for St Helena in January, Nevil dined no less than nine times, but it was not until 1767 that he was elected a member and could dine in his own right, though he dined many times as a visitor before that.

The decision on St Helena had been taken. Now ways and means had to be found to take the Bencoolen decision. Money had been promised by the King, so transport was the next consideration. On 21 July, a memorial was sent to the Admiralty, asking for a man-of-war, informing them of the King's support and suggesting in flattering terms that the Lords Commissioners might 'want no further Sollicitation to give this new Instance of [the Admiralty's] Zeal for the promotion of a Science so intimately connected with the Art of Navigation as well as for the Honour of the Nation.' On 30 July, the Admiralty wrote to say that a ship was to be fitted out for the purpose,[13] which turned out to be the *Seahorse* frigate 24, then at Portsmouth. At the same meeting, Maskelyne submitted an amended estimate of his expenses based on staying one year in St Helena, with three-month passages in an East India Company ship there and back:

Boarding at St. Helena at 6 shillings per day for 1 year	109.10.0
Liquors at five shillings per day for the same time	91. 5.0
Washing at 9 pence per day	13.13.9
Other expenses and incidental charges at 1s 6d per day	27. 7.6
	241.16.3
Liquors on board of Ship for three months going and three months returning	50. 0.0
Total expenses	291.16.3[14]

In addition to the £150 gratuity, the Council said it would pay board at six shillings a day with other expenses as they appeared, providing they did not exceed this estimate.

Several writers have made play on Maskelyne's liquor estimates, and the fact that his assistant's liquor estimate was exactly half his – Weld (1848), "the large sum 'for liquors' will cause some surprise in these temperate times."[14] : Admiral W.H. Smyth (1860), '. . . granted his necessary expenses, in which the sums for drink prove he was not quite what is now ycleped a Tee-totaller.'[15] : while Woolf (1959) quotes Voltaire, 'en ce bas monde, il vaut mieux être gastronome qu'astronome.'[16] It is interesting to compare this estimate with his actual expenses, detailed in Appendix C, which reveal many interesting details. For example, we learn that he took a servant, which must go part way to explaining the liquor discrepancy; and also be took three gallons of lemon juice, which will have protected both of them from the effects of scurvy during the three-month voyage. There is no mention of sickness during the voyages.

On 11 September, the Council made the final decision on the Bencoolen expedition. Mason had said he would be prepared to lead this instead of going as second to Maskelyne, and suggested as his own second observer Jeremiah Dixon, surveyor and amateur astronomer from Cockfield in County Durham. Dixon accepted, and so came together the Mason and Dixon team so well known in North America for their labours in Maryland and Pennsylvania in 1763–7 and for the line that they defined. Mason was replaced as Maskelyne's second observer by Robert Waddington, a mathematics and navigation teacher in Miles Lane, near the Monument,[17] then lodging with Matthew Pigott at Witton near Twickenham, (presumably a relation of Nathaniel Pigott, the York astronomer, who was born at Witton).[18] Maskelyne was directed to order the necessary instruments for Bencoolen, this time from two Fellows of the Royal Society who were astronomers in their own right – two reflecting telescopes from James Short and a clock for twenty guineas from John Ellicott (John Shelton was charging thirty guineas for the St Helena clock.), as well as a pair of globes and a thermometer by Bird. The Earl of Macclesfield lent a quadrant.[19]

After some anxieties as to when the instruments would be delivered, Mason wrote from Portsmouth on 8 December (and again on 19 December) that the Bencoolen instruments and baggage were onboard and that the *Seahorse* was ready, but that the Admiralty had not yet issued sailing orders.[20] On 27 November, Maskelyne and Waddington signed a receipt for the St Helena instruments.[21] Meanwhile, arrangements were being made for them to take passage in the *Prince Henry* East Indiaman. But it was to be January 1761 before either ship sailed.

4

Saint Helena, 1761

The *Seahorse* eventually received her sailing orders and sailed from Spithead (off Portsmouth) for Bencoolen on Friday 9 January 1761. The Council of the Royal Society were therefore somewhat mortified the next week to hear that Dr Morton, one of the Society's Secretaries, had received a letter from Charles Mason, dated Plymouth, Monday 12 January, which started thus: 'I beg the favour you would please to acquaint the Council of the Royal Society that on Saturday last [10 January] at Eleven in the Morng. 34 leagues SW ½ W from Start Point we engaged the L'Grand a thirty-four Gun Frigate, who after an Obstinate Dispute of about one hour and a quarter, Monsieur thought proper to run as fast as possible; after chacing some time in vain, the Captain Steer'd for this Port to refit.'[1]

The *Seahorse* had eleven dead and thirty-seven wounded and there was considerable damage aloft; the instruments were safe though some of the wooden stands were damaged. Mason concluded: 'All our Masts are wounded, and to refit the ship will take up so much time that in my opinion it will be impossible for us to arrive in India in Time to make the observation; and therefore must desire you will please by a line as soon as possible to acquaint me in what manner the Council would please have us proceed.'[2] The prospect of going to Bencoolen now seemed less enticing to Mason and Dixon who, on 25 January, even went so far as to say in a letter to Bradley, 'We will not proceed thither, let the consequence be what it will',[3] suggesting instead they should observe the transit from the eastern Black Sea or from Iskenderun in Turkey. On 31 January, the Admiralty told the Council that they had heard that the gentlemen had absolutely refused to go, and asked, what now?[4]

This was mutiny! That very same day, the Council drafted a stiff letter saying how surprised it was to hear that they declined to go to Bencoolen, pointing out just what the consequences *would* be if they were to abandon the original project, reminding them that they were under contract and had already received money. It threatened legal proceedings. 'That your refusal to proceed upon this voyage, after having so publicly and notoriously engaged in it, will be a reproach to the Nation in general; to the Royal Society in particular; and more especially and fatally to yourselves; and that, after the Crown has been graciously and generously pleased to encourage this undertaking by a grant of money towards carrying it on; and the Lords of the Admiralty to fit out a ship of war, on purpose to carry you to Bencoolen; and after the expectation of this and various other nations has been raised to attend the event of your voyage; your declining it at this critical juncture, when it is too late to supply your Places, cannot fail to bring an indelible scandal upon your character, and probably end in your utter ruin . . . The Council do absolutely and expressly direct and require you to go on board the *Seahorse*, and enter upon the voyage, be the event as it may fall out'[5] They added that, if circumstances (unspecified) made it seem likely they could not get to Bencoolen in time, then the Captain had instructions as to what to do.

Four days later, early on 4 February, the *Seahorse*, this time escorted by the 5th rate *Brilliant* 36 (*Seahorse* was a 6th rate with 24 guns), sailed with Mason and Dixon onboard, duly chastened. They reached the Cape of Good Hope on 27 April where they received the news that the French had captured Bencoolen. Mason reported this – with some pleasure? – in a letter where he said that, as it was too late to try to reach Batavia, the alternative to Bencoolen, it was decided that they should observe the transit at the Cape and that they had landed the instruments. The Dutch were so slow and so few spoke English that it was a great relief when the captain of the *Seahorse* offered the services of the ship's carpenters to erect the observatory and mount the instruments.[6]

Meanwhile, Maskelyne and Waddington were completing their arrangements for passage to St Helena. The *Prince Henry*, East Indiaman, Captain Charles Haggis, having refitted at Blackwall on the River Thames below London, moved down-river to anchor off Tilbury Fort on 18 November 1760. Some time early in December, the astronomers' baggage and instruments were delivered to Botolph's Wharf below London Bridge (see Appendix C), whence they were sent down-river to Tilbury and put on board the *Prince Henry*, who sailed from there on Christmas Day and anchored in the Downs to await the assembly of the next convoy, principally the West India fleet. On 6 January 1761, the Commodore of the convoy gave the order to weigh. The *Prince Henry* anchored in St Helen's Roads off the east end of the Isle of Wight on the 11th and Maskelyne and Waddington, having travelled from London to Portsmouth by road, embarked on the 13th. Meanwhile, more ships were assembling at St Helens.

At 1 p.m. on Sunday 18 January, the convoy weighed anchor and sailed

Saint Helena, 1761

Fig. 4.1. The *Northumberland*, East Indiaman, lying off Jamestown, St Helena. As was common practice in eighteenth-century ship portraits, she is shown from two different aspects, under way and at anchor. It is not certain whether this was E.I.C.S. *Northumberland* of 1780–94, or the ship of the same name of 1800–16, but the *Prince Henry* and *Warwick*, in which Maskelyne sailed to and from St Helena in 1761 and 1762, would have been very similar in appearance. From an oil painting by Thomas Luny, in possession of the National Maritime Museum, Greenwich. (Photo: NMM)

from St Helens, ninety-five sail of merchantmen, mostly bound for the West Indies, escorted by HM Ships *Centaur* 74, *Rippon* 60, and *Port Mahon* 24. The *Prince Henry* parted company from the West India convoy on 1 February when in the latitude of the Canary Isles, proceeding independently to St Helena.[7] Maskelyne began to keep a nautical journal on sailing, recording that he took his departure at 4 p.m. on 21 January (civil date 20 January) when the Lizard bore NE by N, distant 9 leagues.[8]

So Maskelyne first began to gain sea experience and to become involved with practical navigation – in the development of which he was to play such an important rôle in the future. Although no specific directive has survived, he was presumably asked by the Astronomer Royal and members of the Board of Longitude (which included the President of the Royal Society) to try out the lunar-distance method of finding longitude at sea, and particularly Mayer's first manuscript tables, while on passage to and from St Helena, effectively continuing the Board's trials by Captain John Campbell R.N. in 1757–9, already referred to.

In the *Prince Henry*, Maskelyne had with him all that was necessary for lunar-distance observations – a Hadley quadrant by John Bird, of 20 inches radius, with a mahogany frame, brass arc and index, vernier scale and fine-adjustment screw, with the mirrors and shades ground by Dollond; copies of lunar and solar tables by Tobias Mayer and Gael Morris;[9] and the French almanacs *Connaissance des Temps* for 1761 and 1762. In his first report to the Royal Society, he said: 'But my principal attention on board of ship was taken up in observing the distances of the Moon from the Sun & stars with a Quadrant which I had of Mr. Bird in order to be satisfied from my own experience of the practicability of that method of finding the Longitude; and from the near agreement of my observations with one another made in different circumstances I apprehend that a person who will take the pains necessary to make accurate observations and has at the same time leisure and ability to make the requisite calculations will be able to ascertain his Longitude by this method as near as will be in general required.'[10] And leisure was indeed needed, because, at that time, the calculation could take four hours or more. In February and March, he took at least sixteen lunar-distance observations. Although cloudy weather precluded any such observations during the last eight days of the voyage, he reported that the longitude error in his own reckoning on arrival at St Helena on 6 April was only 1½° east, as against the errors of up to 10° of some of the ship's officers.[11] (An error of 1 arc-minute in the apparent distance, whether caused by the observer, the instrument, or the tables, resulted in an error of ½° of longitude.)

The principal town of the island of St Helena is Jamestown on the north-west coast, whose open anchorage is sheltered from the continually blowing South East Trades. The *Prince Henry* anchored there on 6 April 1761 and Maskelyne and Waddington disembarked with their baggage and instruments.

When Halley had been in St Helena a century earlier, he had had very little cooperation and not a little obstruction from the East India Company's then governor, Gregory Field. This time it was quite the reverse: Governor Charles Hutchinson and his colleagues did everything they could to help. In a letter to the Royal Society dated 13 May (sent in the *London*, carrying the news of the capture of Pondicherry on 15 January), Maskelyne describes his actions on arrival:

After a very agreeable voyage of 11 weeks & 2 days we arrived here on the 6th. of April Last. I immediately made it my business to look about the country in order to find a proper place for an observatory & I soon found one which appears to be not only the best but almost the only convenient situation upon the Island for that purpose the valleys being too much block'd up by the high

Saint Helena, 1761

hills which surround them, & the higher hills being almost perpetually covered with Fogs & vapours.

I fix'd upon an intermediate station upon a hill call'd the Alarum Ridge from whence I can see the Sun rise at all times of the year & set at almost all times, my prospect being pretty open every where except towards the South where it is interrupted by a high ridge of Hills; but there I want a view the least. My situation is at some distance below Halley's Mount so call'd here for having been the place of Halley's observations my Astronomical Predecessor. The height of his situation was undoubtedly the reason of the great fogs & vapours which he complains of to have so much disturbed his observations. Perhaps also the Island's being then covered with trees which now is almost in a manner bare of them might be another reason of a greater quantity of cloud & vapours than at present. The Observatory which is now just finish'd [is] a room of 24 foot by 12. I shall be fix'd in a day or two.

The recommendation of the East India Company of me and Mr Waddington to the Care of the Governor Mr Hutcheson & to his assistance in furthering the Observations has been of the greatest service, & without it I do not know what we could have well done; the observatory could certainly never have been finish'd in time if the Governer had not employ'd all the hands that could possibly be spar'd. The great civilities I receive from Mr Hutcheson & the readiness he shews in affording me all the assistance I desire is only what I expected from the Universal excellent character which he bears among all that know him & which I was not unacquainted with in England.[12]

A nice tribute from the 29-year-old astronomer!

Before describing Maskelyne's work on the island, it will help to summarize the scientific tasks he set himself.

(a) The transit of Venus – the primary purpose of the whole expedition, to observe Venus's transit over the Sun's disk from sunrise on 6 June to the moment of her egress, specifically her last *internal contact* with the Sun's limb, some two hours later, using a 2ft focus *Gregorian reflecting telescope* fitted with a *micrometer* for the observation, and Shelton's pendulum clock for the timing.

(b) Checking the going of the clock – finding the moment of local noon by *equal altitude observations* on successive days, using the *astronomical quadrant*, generally called an *equal altitude instrument*.

(c) Latitude and longitude – essential for the transit of Venus. For latitude, by measuring *zenith distances* of heavenly bodies when on the meridian, using the astronomical quadrant. For longitude, by observing and timing the *eclipses of Jupiter's satellites, occultations* of stars by the Moon, and specific moments in lunar and solar eclipses, using a reflecting telescope and clock: if these same events (which do not happen with any regularity) can be observed and timed at, say, Greenwich, then the difference between the local times measured at each place will give the difference in longitude, where 1 hour of time equals 15 degrees of arc.

(d) Parallax of Sirius – by measuring the zenith distance of the star when on the meridian (nearly overhead) with the 10ft *zenith sector* by Jeremiah Sisson, to give a measure of the distance between Sirius and the Sun.

(e) Determining the relative force of gravity compared with that at Greenwich – by measuring the rate of going of the clock on the island (see (b) above) and comparing it with the rate of the same clock with the same pendulum length at Greenwich. The purpose of this was to determine the *Figure of the Earth*; is it oblate (like an orange), or prolate (like a pear) – and by how much?

(f) Determining the moon's parallax in right ascension – when compared with similar observations in other latitudes, this will also help to determine the Figure of the Earth. He used his reflecting telescope on an *equatorial mounting* as a *transit instrument* for these observations, finding the difference in transit times between the Moon and stars having more or less the same *declination*.

(g) Magnetic observations – measurements of *variation* and *dip* with his *variation compass* and *dip circle*. No published results have been found.

(h) Tidal observations – times and heights of high and low water over a period of at least a month in the harbour off James's Fort, using a simple 10ft tide pole.

The transit instrument mentioned in Appendix C was probably not a complete instrument but rather the special mounting for the reflecting telescope mentioned in (f) above.[13]

As we can see from his letter to the Society quoted above, his first thought on landing in early April was to find a suitable place from which to observe the transit, which was to occur two months hence. He knew the troubles Halley had had from the almost perpetual cloudiness at his observatory on the summit of what is today known as Halley's Mount, so he chose a lower site on Alarum Ridge, some two miles or so inland from Jamestown, 'at the Alarum-House' at 1983 feet above sea level,[14] identified in a recent study as the spot height 2130 feet (649 metres) on a 1904 map.[15] Here, the Governor provided the necessary labour to erect a buiding 24ft by 12ft, in which he mounted the clock and equal altitude instrument on massive wooden posts buried in the ground independent of the walls of the observatory, all being completed by mid-May, ready for the transit on 6 June.[16] On arrival, the Governor had provided a house for the astronomers in the valley near James's Fort at an elevation of 85ft,[17] probably on the site known today as 'Sisters' Walk',[18] where he erected his zenith sector and transit instrument and established a meridian mark. When they were working at Alarum Ridge, they stayed at the country house of Matthew Purling, one of the island's Council.[19]

While awaiting the completion of his transit observatory, Maskelyne decided to complete the gravity experiment and temporarily set up the Shelton clock and equal altitude instrument in the Jamestown house. As a result of equal altitude observations between 25 April and 8 May, he decided that the force of gravity at St Helena compared to Greenwich was as 9 975 405 to 10 000 000.[20] So began the astronomical career of a very distinguished and much-travelled clock.[21]

Maskelyne and Waddington were at the observatory on Alarum Ridge before dawn on 6 June and saw Venus silhouetted against the Sun a few minutes after sunrise, though the *limbs* of both Sun and Venus were ill-defined because they were so near the horizon. Then the clouds came up and hid the Sun for an hour. When the Sun shone again, Maskelyne started measuring the distances of Venus from the limb of the Sun, using Dollond's 'curious object-glass micrometer adapted to the reflecting telescope',[22] while Waddington took the passages of the limbs of Venus and the Sun across the horizontal and vertical wires of the equal altitude instrument, both sets of observations being timed by the Shelton clock.

The moment for which they were waiting, the moment for which they had travelled more than five thousand nautical miles – to observe and accurately time the moment of the last internal contact of Venus from the Sun's limb – was approaching. Then, a few minutes before the event was to happen – a cloud covered the Sun! (Quite predictable, said all the old St Helena hands, you were lucky to see anything at all here!)

What a disappointment! However, though the news took some time to reach him, it was some consolation when he learnt that Mason and Dixon had made the observation in good weather at the Cape of Good Hope.

'On this trying occasion he is said to have born his disappointment with so much fortitude as to have said he hoped to meet with better weather to observe the next transit. . .'[23] And so he did: at Greenwich in 1769, he observed the ingress, both external and internal contacts, when the Sun was a little higher in the sky than it was for the egress at St Helena.

Though no less than 120 separate observations of the transit of 1761 were published world-wide by observers from eight different nations, the results were on the whole disappointing. Many observers reported an unexpected phenomenon which came to be called the 'black-drop' effect, when Venus, close to the limb of the Sun, suddenly seems to be joined by a black ligament, making accurate timing of contact very difficult (see Fig.4.2.). Analysis showed all too little consistency in the parallax results, in part due to the black drop, in part to uncertainty in longitude differences. Only twenty-three observed both ingress and egress (where longitude does not matter), the most northerly being in 65°51'N at Tornea in the Gulf of Bothnia, the most southerly in 22°33'N at Calcutta, too small a spread to be significant by themselves. Of the principal French observers outside France, Chappe observed both ingress and egress in Tobolsk in Siberia, Pingré egress on Rodriguez I. in the south Indian Ocean: Le Gentil never reached Pondicherry and Boscovich was also frustrated. The only other British observer of any importance outside Britain was John Winthrop of Harvard College who successfully observed the egress in Newfoundland. But all were agreed that this had proved a most useful exercise from which lessons could be learned which would make a success of the 1768 transit.[24]

Though Maskelyne was to remain on the island for another seven months to continue his observations on the parallax of Sirius, Waddington had been hired for the transit only. He embarked in the *Oxford* East Indiaman and sailed for England on 29 June in company with seven other ships. While on passage, he took many lunar-distance observations and so impressed Captain Webber that he refused to accept payment for Waddington's passage.[25] After a very rough voyage, in which Portland Bill was the first land sighted on the way up-Channel, the *Oxford* eventually came to anchor off Deal Castle on 15 September.[26] After hearing Waddington's report, the Royal Society's Council wrote a fulsome letter of thanks to the East India Company for all the kindness shown and help given by Governor Hutchinson and the Council in St Helena.[27]

After his disappointment with the transit, Maskelyne turned his atten-

Saint Helena, 1761

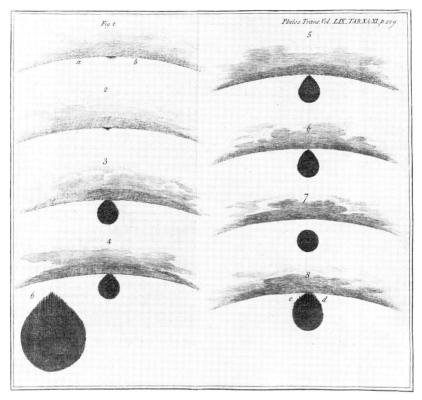

Fig. 4.2. The Transit of Venus, 1769. The first external contact of Venus's disc with the Sun's limb (1), and first internal contact (3), as seen by Rev. William Hirst with a 2-foot reflector at the Royal Observatory, Greenwich. From *Phil. Trans. R. Soc.* 59, plates 10 and 11. (Photo: RGO)

tion to the zenith observations of Sirius. The sector had been set up in Jamestown and he had made his first observation there on 19 May. By the 25th, things had begun to go wrong: the results were inconsistent. He thought that the object glass was at fault and ordered another from England in June. Up to the end of August, he took many observations for longitude on Alarum Ridge, not having been able to see any eclipses of Jupiter's satellites before the transit.

September brought very bad weather, so he moved the sector from Jamestown and set up another observatory near Sandy Bay on the other side of the island. The measurements taken with the zenith sector depended upon a long plumb-line suspended from the upper pivot of the 10-foot-long telescope, being read off against a brass scale at the lower end: this defined the vertical and the zenith point from which the zenith distances of the stars on the meridian were read off (see Fig.4.3).

36 *Nevil Maskelyne – the seaman's astronomer*

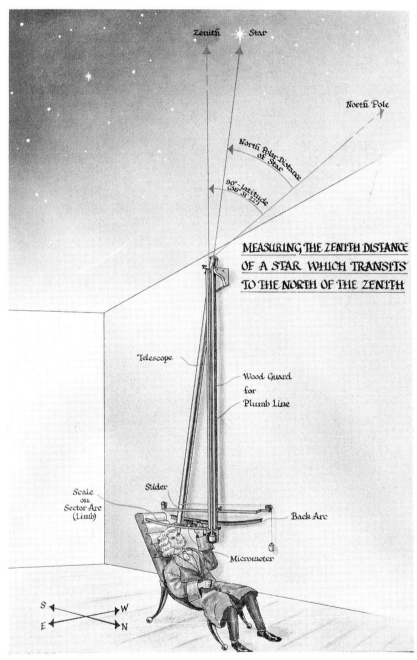

Fig. 4.3. Measuring the Zenith Distance of a Star with a Zenith Sector. This shows Bradley's 12½-foot zenith sector at Greenwich, but the 10-foot sector which Maskelyne used at St Helena and Schichallion was exactly similar in principle.   (Photo: NMM)

Saint Helena, 1761 37

However, as will be discussed in detail in Chapter 12, where one was using such a sector near a mountain – and St Helena itself could be considered as such – Newton's gravitational theory said that the plumb-bob would be attracted towards that mountain – only by a very small amount it is true, a few arc-seconds, but enough to upset the precision of *absolute* measurements of zenith distance. (Providing it does not change, such an error in absolute zenith distance need not affect stellar parallax measurements, because they depend upon the *difference* between zenith distances measured six months apart.) The reason Maskelyne moved to the other side of the island was to see whether he could detect such an attraction. On the north-west coast, the plumb-bob should be attracted to the south and the zenith distances measured should be too great in a northerly direction: on the south-east coast, the reverse should occur: a mean of measurements at each place ought to yield the true zenith.

In the event in St Helena, instrumental errors entirely swamped the effect of what came to be called the Attraction of Mountains. However, many years later, Maskelyne himself did manage to measure this effect on the mountain of Schiehallion in Scotland, using the very same sector and clock – but that story will be told in Chapter 12.

After moving the sector out of his house in Jamestown, he abandoned his cloud-ridden Alarum Ridge observatory, where he had begun his observations for lunar parallax (see (f) above) on 3 September. On 24 September the Shelton clock was fixed up in the upper room of his house in Jamestown, and the equal altitude instrument in a special observatory at ground level. Here, he continued his lunar parallax observations.[28]

On 16 October, Maskelyne was delighted to welcome some astronomical company when Mason and Dixon arrived from the Cape with their instruments. After some discussion, it was agreed that Dixon should return thither, taking with him the Shelton clock, so that he could repeat with that clock the gravity-measurement experiments (see (e) above) which he had already done with the Ellicott clock. In that way, results could be obtained from both clocks in both places. The Shelton clock was accordingly taken down and packed up for transport, and a journeyman clock by Shelton (which swings seconds, has a loud tick, and strikes every minute, presumably brought from Greenwich by Mason) was erected in its place. Mason fixed the Ellicott clock in the lower observatory.

While awaiting Dixon's return, Mason assisted Maskelyne not only in his astronomical observations, but also in tidal observations which involved regular inspection of the 10ft tide pole erected in front of James's Fort, day and night from 12 November to 22 December 1761.[29]

Just before Christmas, Maskelyne and Mason moved into the country to continue observations with the sector.³⁰ It was probably at that time that there occurred the episode of the German soldier, the first of several anecdotes which prove Nevil Maskelyne to have been a man of great compassion, always ready to help his fellow men. While walking near the fort at Sandy Bay, he met a German soldier 'who appeared by his behaviour & address to be above the common sort.' He said that, while destitute in London, he had been induced to sign up for the Company's service. However, things did not turn out as he had been led to believe, and the one thing he now wanted was to return to England and thence to Germany 'where he could get into a better situation in life.' Maskelyne went to see the Governor and himself paid the levy money for a substitute so that the German could return home.³¹ The two mountain massifs on either side of Sandy Bay are known respectively as the Devil's Garden and the Gates of Chaos: perhaps our German friend found the country unwelcoming.

Dixon and the Shelton clock returned on 30 December 1761 and he and Mason embarked for England shortly after, leaving Maskelyne still trying to find the cause of his abortive observations with the sector, which had by now returned to Jamestown. By February 1762, he had decided that the fault lay in the method of suspension of the plumb-line, and that that could not be rectified in St Helena. He accordingly packed up his instruments and baggage and sailed for England in the *Warwick* East Indiaman, Captain James Dewar, on 19 February 1762, escorted by the *Terpsichore* 24, in which Governor Hutchinson took passage.

Though the ten months Maskelyne spent ashore in St Helena had been frustrating in the extreme as far as astronomical results were concerned, they certainly gave him a wealth of practical experience which was to stand him in good stead in his future career. And now he was to gain more practical experience, this time afloat, in lunar observations at sea. He felt that he had proved the utility of Mayer's tables and of the method generally on the outward voyage. Homeward bound, therefore, he took few observations himself in the first month or so. What he did do, though, was to encourage the ship's officers to take such observations and to show them how to do the laborious calculations – which was to pay great rewards in the future.

From the end of April, approaching the Chops of the Channel, he started taking as many observations as possible. The results promised well as observations taken with different stars on the same day agreed with each other and these agreed from day to day. Then on 9 May, the Scilly Isles were sighted and a lunar observation taken that same day proved to be only twelve nautical miles in error. All the way up the Channel he checked his longitude observations with land bearings, placing the Lizard's longitude within 0°50′ of its assigned place. 'Longitude', he claimed, 'can always be found within a degree, or very little more, which answers to about 40 geographical miles in the latitude of the English Channel.'[32]

Maskelyne took his last observation at 9 a.m. on Friday 14 May 1762, when Eddystone Light bore N22°E, 3 miles. The *Warwick* and the *Terpsichore* anchored in Plymouth Sound for the night, moving up to anchor in the Hamoaze the next day, whence Maskelyne travelled to London by coach.[33]

5

The Barbados trials, 1763–4

While Maskelyne was in St. Helena, the Board of Longitude had been very much concerned with the work of the clockmaker John Harrison. On 18 July 1760, twenty-three years after his first appearance at a meeting of the Board of Longitude, he had declared himself ready to make the trial with his third large timekeeper – to the West Indies as demanded by the 1714 Act. But he also produced to the meeting a large watch, today known as H4, which, he said, answered beyond his expectation, though it still needed some adjustment for winter temperatures. He suggested a trial be planned for April 1761. The Board gave Harrison yet another £500 – making a grand total of £3 250 since 1737 – to finish the adjusting, so that both the large timekeeper and the watch could be tried together.

On 12 March 1761, the Board convened again, having been informed by Harrison that H4 was ready and that, in view of the fact that he was 68 years of age, his son, William, would conduct the trial on his behalf. William and the two timekeepers proceeded to Portsmouth to await a ship, which it was hoped would be the *Dorsetshire* 70, commanded by Captain John Campbell – someone who knew more than anyone else in the Royal Navy about the longitude problem at sea. This plan was frustrated when the *Dorsetshire* was ordered to the Mediterranean and many months elapsed before sea trials were possible.

In the event, Harrison decided to stake his all on the watch and to abandon the large timekeeper. On 18 November 1761, William Harrison, accompanied by John Robison, an astronomer appointed by the Board, sailed in the *Deptford* 60 for Jamaica with H4 onboard, where they arrived on 19 January 1762. Robison, who had been recommended by Admiral Charles Knowles, a member of the Board of Longitude, had carried out surveys in Canada under General Wolfe while tutor to Knowles's son and later became Professor of Experimental Philosophy at Edinburgh and first Secretary of the Royal Society of Edinburgh. Because of wartime pressures, Harrison and Robison were unable to take more than one equal altitude observation in Jamaica to check the going of the watch before having to embark in the *Merlin* sloop for passage home on 28 January, reaching Spithead after a very rough passage on 27 March 1762. The Harrisons claimed that, in the 81-day voyage from Portsmouth to Jamaica, taking into account how much they predicted it would gain or lose each day based on a few days' observations at Portsmouth (but which they had failed to declare to the Board before sailing), H4 lost only 5.1 seconds, seemingly an absolutely incredible result, more than qualifying for the £20 000 offered by the Longitude Act.

But the Board thought otherwise, refusing to accept Harrison's claims from the Jamaica trial. At their meeting of 17 August 1762, they resolved that 'the experiments already made of the watch have not been sufficient to determine the Longitude at Sea'. Nevertheless, they found the invention 'of considerable Utility to the Publick', so they awarded him £1 500 immediately with the promise of an extra £1 000 after a further trial to the West Indies, with the proviso that £2 500 should be deducted from any future reward and that the winning timekeeper should become the property of the nation.[1]

Meanwhile, there were other developments on the longitude front. Tobias Mayer had died in Göttingen at the early age of 39 on 20 February 1762. Shortly before this, he had received copies of lunar-distance observations taken at sea by a Danish ex-pupil Carsten Niebuhr.[2] These pleased him so much that he asked his wife to send them to England with a copy of his lunar and solar tables considerably improved by minor corrections made in the seven years since his first tables had been sent to Lord Anson. His widow's memorial dated 13 June 1762 was considered at successive meetings of the Board of Longitude from 17 August 1762, though, as we shall see, no action was taken until 4 August 1763. Meanwhile, the Peace of Paris had been ratified on 5 May, so Britain was no longer at war with France.

Another loss was the Abbé Nicolas-Louis de Lacaille, who died in Paris at the age of 48 on 21 March 1762. He had himself made lunar observations on his way home from the Cape of Good Hope by way of Mauritius and Réunion in 1755, though the lunar tables he used were less precise than those of Mayer. In the *Connaissance des Temps* for 1761 (published 1759) Lalande had published Lacaille's explanation of his method, together with diagrams for graphical solutions and pre-computed lunar distances every 4 hours for the month of July 1761.[3] It was Lacaille's method of working (though not his diagrams) which Maskelyne was to recommend in the *British Mariners' Guide*, and the pre-computed lunar-distance tables he eventually published in successive editions of the *Nautical Almanac* from 1767 were also based on Lacaille's example.

Then, on 17 July 1762, Bradley died, being succeeded as Astronomer Royal on August 12 by his Oxford colleague, Nathaniel Bliss, Savilian Professor of Geometry and friend of the President of the Royal Society, the Earl of Macclesfield.

Maskelyne returned to Chipping Barnet to his lodgings with Richard Hassell and to his pastoral duties in May 1762. On 27 May, he reported to the Council of the Royal Society and made arrangements for the defects in the suspension of the zenith sector plumb-line to be corrected by the maker Jeremiah Sisson. On 23 June, he reported to the Court of Directors of the East India Company, thanking them for all the help the company had afforded, and pointing out 'a method for the more exact discovery of the Longitude at Sea, which thro' his Assistance had been practiced by the officers of the ship *Warwick* in her homeward bound passage: and intimating that if the said Method is recommended by the Company to the Commanders of their ships it may be of great Utility in Navigation.'[4] Chairman Thomas Rous, a friend of Clive's, thanked him and he retired, saying he would be happy to be of any further service to the Company.

It must have been soon after this that he started writing his *The British MARINER'S GUIDE containing Complete and Easy Instructions for the Discovery of the LONGITUDE at Sea and Land, within a Degree, by Observations of the Distance of the Moon from the Sun and Stars, taken with HADLEY'S Quadrant*. This contained an English edition of Mayer's first manuscript tables and gave simple instructions for making lunar-distance observations and computing the longitude from them. He gave a worked example of actual observations taken in the *Warwick* when within sight of the Lizard at about 2 in the morning on 23 May 1762, when a lunar distance between the star Antares and the Moon gave a longitude of the Lizard which he claimed was within 50 arc-minutes of longitude (32 nautical miles) of its accepted place. Eventually published in April 1763, one wonders how much the advocacy of the method to the Company and the writing of the book were stimulated by possible awards under the Longitude Act – though he certainly never submitted any claims.

Waddington also burst into print, a copy of his book on the same subject, dedicated to George Grenville, First Lord of the Admiralty,[5] being presented to the Royal Society two months before Maskelyne's. This contained lunar tables for 1763, while a supplement containing tables for 1764 was published in January 1764.[6] Tables for 1765 and subsequent years were promised but do not seem to have been published. (He became Second Mathematical Master of the Royal Naval Academy at Portsmouth, but was dismissed in 1766,[7] returning to London where he published *An Epitome of Theoretical and Practical Navigation* in 1777, and an appendix entitled *The Sea Officer's Companion* in 1778.)

Of Maskelyne's activities between June 1762 and June 1763, we know little except that he was busy with his book and was still curate in Chipping Barnet. That he assiduously attended Royal Society meetings can be inferred from the dinner register of the Mitre Club where he is recorded as dining as a visitor ten times during that period.[8] He was an unsuccessful candidate for the Woodwardian Professorship of Geology at Cambridge before December.[9] None of the early occupants of this chair were full-time geologists: the successful candidate was his friend John Michell, some eight years his senior.

Then, at a meeting of the Royal Society on 9 June 1763, he presented a long address (taking up nearly eight pages in the Journal Book[10]) asserting that 'by an unhappy concealment & retention of the Observations made at the Royal Observatory, the improvements of Astronomy have been kept 50 years behind hand.' This had been brought to a head because none of Bradley's twenty years' worth of Greenwich observations had ever been published or made available to those who so urgently needed access to them. Now his executors were threatening to seize his papers – and actually did so two years later – arguing they were private property.

Maskelyne pointed out how much astronomy and navigation were being hindered world-wide because previous Astronomers Royal claimed that the observations made at the Royal Observatory were their own private property. For example, Flamsteed's insistence on withholding his observations 'not only from the publick, but even from his learned friends, a long while deprived Newton of the means of his establishing his theory of the moon, & might have done so longer, if that great Man, in conjunction with other learned ornaments of this Royal Society, had not obtained an order from superior authority, for selecting & publishing such of the Observations made at the Royal Observatory as they should think proper.' Flamsteed's observations were eventually published in

1725 with his British Catalogue of stars, but, as might be expected, errors had since been found so that today, 'to the great discredit & dishonour of this nation, this Society, & to the general state of learning in this Kingdom', English astronomers had to use the product of *French* observations – Lacaille's *Fundamenta Astronomiae*, a star catalogue published in 1757 – because they could not obtain those made in their own Royal Observatory.

In Maskelyne's own case, Bradley had refused to let him see any of his observations while he, Maskelyne, was formulating his proposals for the measurement of the annual parallax of Sirius in St Helena. Since his return from that island, Bliss (who was now the temporary custodian) had sent him extracts from Bradley's observations which indicated that Sirius's annual parallax would be hardly measurable, whereas he had based his plans on Lacaille's estimate of 9 arc-seconds. (Today's figure is $0''.375$) He continued:

If therefore I had but had the same opportunity of inferring the little probability of an annual parallax of Sirius, from the Greenwich Observations, before I went to St Helena, as I have had since my return, the expence of the ten foot Sector might have been saved to the Society, & I myself should have escaped a disagreeable stay in that island. . .

Maskelyne then proposed (a) that as permanent Visitors to the Royal Observatory, the Society should demand the astronomical results from the present custodians so that they could be published; and (b) that in future the Astronomer Royal should be required to deliver his observations to the Society annually for publication. On motions from Taylor White, barrister and secretary of the Foundling Hospital (of which William Harrison was a Governor), the Society agreed with Maskelyne's proposals in principle but directed the Council to ascertain the legal position. The saga of the publication of Bradley's observations will be continued in Chapter 10.

But now, Harrison began to realise that there was another threat. An account of Maskelyne's successful use of Mayer's tables to and from St Helena had been read to the Royal Society in June 1762, only a month after his return,[11] and he was known to publishing a book on the subject. The Longitude Act stipulated that only the *first* proposal for a practicable and useful method of finding longitude would receive the main award. Perhaps some protagonist of the rival lunar-distance method – perhaps even Maskelyne himself – might win the main prize before he, Harrison, was ready. On the Board, Macclesfield and Bliss were known to favour the lunar method.[12]

Early in February 1763, a pamphlet supporting Harrison was sent to all Members of Parliament. Entitled *An Account of the Proceedings, in order to the Discovery of the Longitude at Sea, in a Letter to the Right Honourable ******, Member of Parliament*, the author was stated to be 'A member of the Royal Society'. The published letter was followed by appendices reproducing documents of evidence. This pamphlet is believed to have been the joint work of two of Harrison's friends in the Royal Society: Taylor White, already mentioned, and James Short, the optician and astronomer who had made Maskelyne's telescopes for St Helena.[13]

On 26 February, having thus prepared the ground, Harrison received the Board's approval to petition Parliament. On 2 March, Chancellor of the Exchequer Sir Francis Dashwood presented to the House of Commons a memorial from Harrison, in which he asked Parliament to grant him such part of the reward mentioned in the Act of Queen Anne as they thought adequate now, and the remainder when the utility of his method – the watch – should be proved. In return, he would immediately disclose 'the Manner and Principles for forming it' to whomsoever Parliament might appoint, so that other workmen could make such watches. But he asked that no other longitude proposals concerned with timekeepers should be considered until judgement had been passed on H4, following a further sea trial. The matter was referred to a Committee which presented its report on 12 March, having examined William Harrison, Bevis, Short, George Lewis Scott FRS (a mathematician), and Robert Bishop (a pilot).[14]

As a result, the Act 3 George III c.14 – 'for the Encouragement of John Harrison, to publish and make known his Invention of a Machine or Watch, for the Discovery of the Longitude at Sea' – was pressed through with some speed and became law on 31 March 1763. Parliament welcomed Harrison's offer to 'discover' his secrets, in view of the fact that 'his present advanced Age, the Weakness of his Sight, and the Danger of the voyage to the Health and Life of his Son, upon whom alone the Success of the said Invention will depend, may risk the Loss of so useful a Discovery, not only to the Disadvantage of the said John Harrison, but also to the very great Detriment of Mankind.'[15] The Act nominated a committee of eleven – three Vice Presidents and two ordinary Fellows of the Royal Society (including James Short), a professor from Cambridge (John Michell), and five watchmakers[16] – to receive Harrison's disclosures, after which he was to receive another £5 000. It supported Harrison's rights to the invention – but only by means of a timekeeper (so he was not protected against lunar-distance proposals) – provided the final decision was made within four years.

Even before the passing of the Act, the Duc de Nivernais, who was in London negotiating peace terms, wrote to his superior in Paris saying that he had been told that the examination of H4 would take place in public and suggesting some suitable Frenchman should be sent to attend. As a result, the Académie des Sciences sent Camus, an astronomer, and Berthoud, a clockmaker, to join Joseph-Jérôme Lefrançais de Lalande, editor of the *Connaissance des Temps*, who was already in London.[17] In fact, it was never Parliament's intention that the first examination should be in public, although it *was* their intention 'that this Machine should be made publick for the common benefit of Mankind' and that, when the disclosure had been made, they would 'publish it to the world.'[18]

When Harrison saw the text of the Act, he must have been pleased. It gave him everything he had asked for, with the added bonus of £5 000 as soon as he had made the 'discovery' to the commission of eleven. What he had not bargained for, however, was the commission's interpretation of what should constitute 'discovery'. The Act said that he, Harrison, his executors or administrators, should:

make or cause to be made a full and clear Discovery of the Principles of his said Instrument or Watch for Discovery of the Longitude, and of the true Manner and Method in which the same is and may be constructed. . .

and added that the members of the commission and Harrison himself were:

hereby required to publish and make the same known, so that other Workmen may be enabled to make other such Instruments or Watches for the same purpose. . .[19]

When the Commission met on 12 April 1763 to consider the Act, they interpreted it (James Short dissenting) as requiring complete drawings to be made and H4 to be dismantled before the Commission so that the purpose of each part could be explained, and also that two or more copies of the watch must be made by other workmen under Harrison's superintendence, to be tested before the final decision was made. Harrison was horrified. How could he, already 70, prepare H4 for trial within a reasonable time if the watch had to be completely dismantled first? And to make two copies as well would delay things even more. No, he would not disclose his secrets under those terms. He was prepared to forgo the immediate £5 000 (which in any event was to be deducted from the main prize if he got it eventually) and said he would stake his all on a second sea trial.

And it is here that Nevil Maskelyne re-enters the story. Although we have no direct evidence, we can be pretty sure that, despite his pastoral duties in Chipping Barnet, he managed to find time to keep abreast of developments in the Harrison affair. He also met Lalande (with whom he was so frequently to correspond in the future on Nautical Almanac matters) for the first time, when they were both guests at the Mitre Club on 17 March, and again on 12 May, when Camus, La Condamine and Berthoud were also guests. Of the first dinner, Lalande in his diary says that ordinarily the cost of dinner was three shillings and one sol (penny?), but that on this occasion, because of claret at five shillings a bottle, it was four and a half shillings. Of the food, he mentions only 'plomb poutingin' and 'marque potinger, etc.'[20], but the Treasurer records the following menu in the dinner register:

Codshead & Whiting	Fresh salmon
2 dishes Boiled Fowles	Attam
Tongue & udder	Calves head hached
Turkey Roast	Clump of Beef
Fricase of Lamb	2 dishes Hot Lobsters
Coast of Lamb & Mint	2 dishes Minced pyes
2 boiled puddings	2 apple pyes
Marrow pudding	Butter & Cheese

Eleven members dined, with ten visitors, and the food cost 2s.6d each.[21] The diary also discloses that Maskelyne and Lalande met on at least five other occasions, generally for a meal, sometimes at Edmund Maskelyne's lodging in Maddox Street off New Bond Street.[22]

The Board of Longitude met on Thursday 4 August to arrange the second sea trial of H4, Captain Campbell attending at William Harrison's request. First, the technical arrangements for the trial were discussed, to try to obviate the defects which beset the first trial. First, it was agreed that the Harrisons should declare the rate of going of the watch immediately before sailing, in a sealed document sent to the Secretary of the Admiralty. Secondly, against Harrison's wishes, it was decided that the difference of longitude between the two places should be decided by corresponding observations of the eclipses of Jupiter's satellites, the same events being observed simultaneously in the West Indies and in England. At the same meeting, the Board decided to take the opportunity of giving a sea trial to two other longitude proposals: Mayer's last manuscript lunar and solar tables; and Irwin's marine chair, of which more anon.

At meetings the following Tuesday and on 5 September, it was arranged that the necessary observations – both the equal altitudes of the

Sun before and after noon for checking the going of the watch, and Jupiter's satellites for ascertaining the longitude difference – should be taken in Portsmouth by John Bradley, the late Astronomer Royal's nephew and Campbell's purser in the *Dorsetshire*; while Nevil Maskelyne and Charles Green, Bliss's assistant at Greenwich, agreed to go to Barbados (instead of the fever-ridden Jamaica, at Maskelyne's request), the former as a naval Chaplain with a gratuity of £300 (the Royal Society had only given him £150 for St Helena), the latter to be paid as a Purser of a 5th Rate, both plus expenses. Robison, who had been the astronomer on the Jamaica trial, withdrew at the last minute for family reasons.[23]

The Barbados expedition

Having put his affairs in order and having given Power of Attorney to his brother Edmund, who was staying with their Uncle Nevil at Purton Down,[24] Nevil set off for Portsmouth on 9 September. Some time in the next few days, he and Green joined H.M.S. *Princess Louisa*, a 4th-rate of 60 guns, Captain Joseph Norwood, while at anchor in Spithead. She was wearing the flag of Rear Admiral Richard Tyrrell, going out to command the Antigua Station.

The log of Lieutenant Patrick Fotheringham for 13 September 1763 records the following:

Fresh gales & squally. Came alongside a Hoy with two Marine Chairs and apparatus for observing the Planet Jupiter in order to finding ye Longde. at Sea the Commissioners for ye Discovery (*sic*) to examine these Machines under ye Direction of Adml. Tyrrell in ye course of his Voyage; Do. came on Bd. Mr. Christopher Erwin the Inventor of ye Marine Machine.[25]

The purpose of the chair was to provide a steady platform for observing Jupiter's satellites and occultations of stars by the Moon through a telescope. Earlier models had been tried in the *Magnamine* 74 and other naval ships in 1759 and the Board of Longitude had granted Irwin (not Erwin) £500 and had arranged the present trial. Lalande had seen one of Irwin's marine chairs in Jeremiah Sisson's workshop in the Strand on 29 March.[26]

The *Princess Louisa* moved to St Helens Roads on 18 September, sailing for trials on the 20th, returning to St Helens to embark a new capstan on the 22nd, sailing finally for Barbados on the 23rd. On the 25th, she was joined by the *Milford*, a 6th rate of 28 guns. On October 1, the *Milford's* mast went by the board and for the remaining five weeks of the voyage the *Princess Louisa* had her in tow.

Maskelyne had been instructed to take lunar observations at sea to try out Mayer's last manuscript tables and the Board had lent him a brass sextant by John Bird. He had also his own mahogany quadrant by Bird, a watch (not a chronometer) by Ellicott, and Admiral Tyrrell's excellent pocket watch by Grignon.[27] He and Green took many lunar-distance observations during the voyage and he claimed that the results of his last observation before reaching Barbados were within half a degree of the truth.[28]

'My friend Irwin's machine proves a mere bauble, not in the least useful for the purpose intended.'[29] This was Maskelyne's private comment, which he put into more formal language in his report to the Board, in which opinion he was supported by both Tyrrell and Green. On 8 October, the first clear night when an eclipse of one of Jupiter's satellites occurred, Maskelyne found that, even in a flat calm, the planet and the satellites moved so rapidly to and fro across the field of the telescope that no precise timing of the eclipse was possible. Green found the same thing on many occasions but did manage to get one result in the chair on October 23, though Maskelyne claimed to have done just as well at the same time with a telescope on deck.[30] But his final conclusion was that the eclipses of Jupiter's satellites could never be satisfactorily used to find longitude *at sea*, because the telescope magnification of 50 times, essential for observing such eclipses, was far too high for use in a moving ship.[31]

Having cast off the tow, the *Princess Louisa* anchored in Carlisle Bay off Bridgetown, Barbados, at noon on Monday 7 November 1763. The following Thursday, Maskelyne and Green landed with their instruments – the Shelton clock borrowed from the Royal Society, the Board's 2-foot focus 3.8-inch aperture Gregorian reflecting telescope, an equal altitude instrument, and Maskelyne's own 18-inch Gregorian, all by Bird. The next day, so that no time should be lost while they had no proper observatory, they set up the clock at Willoughby Fort, adjusting the pendulum to the same length it had been in St Helena. On Saturday, they began observing equal altitudes for setting the clock and Jupiter's satellite observations for settling longitude.[32]

Maskelyne found Barbados very pleasant. After telling brother Edmund in a letter of 29 December that his observations at sea were sometimes rather too fatiguing, he continued:

Since my arrival here I have passed my time much more agreably, to which the great civilities I have received from the gentlemen of this place have not a little contributed. . . This country is much better adapted for celestial observations than England, the Air being generally much purer & serener, insomuch that for this month past I have miss'd scarce any observations that occurr'd.

The Heat here is indeed at times somewhat too great for an European Constitution. However I have hitherto escaped without receiving any injury from it; tho' I have had some little disorders, which are now pass'd off, & will I hope season me to the climate.

The face of the country here is very pleasant in general, & as green if not greener than in England, but not with Grass, but with India Corn, Guinea Corn, & Sugar cane plants; the first of which is chiefly used for sustaining the poultry & sometimes also the cattle, & the second is the chief food for the negroes.

I lodged here at Mr. Knight's an attorney a gentleman to whom I was introduced by one of my shipmates & who very politely desired me to make use of his house, where I am very agreably situated.[33]

At Willoughby Fort on the water's edge they had to observe in the open air. Maskelyne chose, as the site for the main observatory, Constitution Hill just outside Bridgetown, 200 yards east of St Michael's Church, 'by a gentle ascent from the town, open & airy which are no mean recommendations of a situation in this hot climate', as he said in his report to the Board, continuing:

The Observatory which is now erected, tho' not quite finished, is a wooden building 8 foot long & 12 wide, having also a little room or projection of 6 foot square on the South side for containing the equal altitude instrument. The clock I propose to place on the East side at a little distance from the South end. I have two shutters in the roof at the South end which when opened will enable me to observe the celestial Phenomena at considerable elevations, or when they pass near the zenith.[34]

Observations in the new observatory started on 6 January 1764.

Back in Portsmouth, John Bradley and William Harrison checked the going of H4 during March, by equal altitudes and by comparison with a Shelton clock borrowed from the Duke of Richmond from nearby Goodwood. On 26 March, Harrison submitted his official declaration of the predicted rate of the watch – one second a day gaining throughout the voyage. On 28 March, Harrison and H4 sailed in the *Tartar* 28, Captain Sir John Lindsay, reaching Bridgetown on 13 May.

On arrival, Harrison landed to seek out Maskelyne and Green. It was then that an unpleasant dispute arose, mentioned in Harrison's manuscript *Journal* but nowhere else. This is how William Harrison describes it:

On Mr Harrisons landing at Barbadoes he was told that Mr Maskelyne was a Candidate for the Premium for discovering the Longitude and therefore they thought it was very odd that he should be sent to make the Observatuons to Judge another Scheme Mr Maskelyne having declared in a very Public manner that he had found the Longitude himself and he had also shewn a letter from a friend who said he was very sorry the Commissioners should give him the trouble of this second voyage before they gave him the reward therefore it was plane from this that Mr Maskelyne's friends were well acquainted with what intention Mr Maskelyne went to Barbadoes.[35]

Harrison told Captain Lindsay about this. Early on Monday morning, they both took the watch to the observatory where Harrison told Maskelyne what he had heard, saying that he considered Maskelyne a most improper person to make the equal altitude observations which would be the main evidence for or against the watch. Whether or not Maskelyne denied the allegation we do not know, but he was greatly upset by this slur on his character and insisted that he must take the observations as ordered by the Board. Presumably as a result of Captain Lindsay's arbitration, a compromise was reached that Maskelyne and Green should observe on alternate days. And so it was: Maskelyne observed equal altitudes of the Sun before and after noon on Monday and Wednesday, Green on Tuesday and Thursday, all in the presence of Captain Lindsay and three other witnesses.

Concerning this incident, Harrison's biographer, Colonel Quill, has this to say:

It is difficult to understand why Harrison should have waited until his arrival at Barbados to raise this objection. Both he and his father had known as far back as the Board meeting of 4 August previously that the observations at Barbados were to be taken by Maskelyne and Green, and in addition there had never been any secret concerning the former's interest and support of the lunar method. Indeed Maskelyne had received specific instructions from the Board that during the voyage to and from Barbados he was to 'make observations of the Moon's motions, and to try the accuracy of the said [Mayer's last manuscript] tables.'[36] As it so happened, these tables had enabled Maskelyne to predict the longitude of Barbados to within ½°, so it was only natural that after landing at Bridgtown he should have shown enthusiasm at the possibilities opened up by the tables as a solution to the longitude problem. The objections raised by William Harrison reveal the spirit of suspicion and antagonism against Maskelyne that seems to have been continually in his mind, an attitude which was shared by his father, and which was to persist to the end.[37]

Just what social contact these two had between this incident and Harrison leaving Barbados a fortnight later is not on record, but, on one side at least, the rift cannot have been too great in that Maskelyne was one of twenty-three who proposed William Harrison for Fellowship of the Royal Society the following February.

Harrison, Green and H4 sailed for home in the *New Elizabeth*, merchantman, Captain Robert Manley, on 4 June 1764, disembarking at Surrey Steps, London, on 18 July. Maskelyne stayed to complete his longitude observations and to make observations for the Moon's parallax in right ascension, in continuation of those taken with the same telescope in St Helena. (See p.32 above.)

Maskelyne sailed from Bridgetown in the *Britannia*, merchantman, Captain Hesketh Davis, on August 30, arriving in England after a six-week voyage on 12 October 1764.[38] Once more, his landfall proved lunar observations to be highly satisfactory, the Isle of Wight being sighted within 11 miles of the predicted position.[39]

6

Astronomer Royal, 1765

After a six-week voyage from Barbados, Maskelyne arrived in London on the evening of Friday 12 October 1764, going to the house of his cousin John Walsh in Chesterfield Street. Calling on his sister Margaret, now Lady Clive, in nearby Berkeley Square (Clive himself had sailed for Bengal in June, without his family but taking with him Nevil's brother Edmund as ADC), Maskelyne learnt some – to him – exciting news: Bliss the Astronomer Royal had died on 2 September and Maskelyne's Cambridge friend Anthony Shepherd, the Plumian Professor and member of the Board of Longitude, had told Lady Clive that her brother was a strong candidate for the vacant post, having in particular the backing of the Earl of Morton, who had become President of the Royal Society on the death of the Earl of Macclesfield in March.[1] Gould says that Bliss was already dying of consumption when he was appointed Astronomer Royal,[2] so perhaps Maskelyne was not entirely taken by surprise.

'I flatter myself', wrote Maskelyne to Shepherd on 20 October, 'with the hopes of obtaining the Professorship at last which you know was always my earnest wish.'[3] And he set to work trying to make certain this came about.

As the name implies, the post of Astronomer Royal – the Royal Professorship at Greenwich – was in the gift of the King himself, advised by his ministers and by the scientific establishment led by the President of the Royal Society. Of these advisers, Maskelyne lost no time in going to the top. He had arrived in London on Friday: between his first letter to Shepherd on Monday and his second on Saturday, he had called on the Prime Minister (the Hon. George Grenville), the Secretary of State (the Earl of Sandwich), the First Lord of the Admiralty (the Earl of Egmont), and the President of the Royal Society (the Earl of Morton), all of whom at one time or another had been members of the Board of Longitude and knew his work. Grenville and Sandwich had already recommended him to the King. Morton had also been 'greatly my friend, and has recommended me to the Ministry in the strongest manner, so long as 4 months ago, before Bliss's death.'[4]

The contest, he said, seemed to lie between himself (aged 32) and the 46-year-old Rev. Joseph Betts, who had already been nominated Bliss's successor to the Savilian Chair of Geometry at Oxford. The academic establishment at Oxford, led by the Chancellor, George Henry Lee, third Earl of Lichfield, strongly recommended that Betts should also be Bliss's successor at Greenwich. After all, the last three Astronomers Royal – Halley, Bradley and Bliss – had all been Oxford men. Surely tradition must be followed!

But Cambridge was not to be outdone. The Prime Minister (who had matriculated from Christ Church, Oxford, in 1730), told him that Betts had already obtained written testimonials from Oxford. Maskelyne should get testimonials from his own learned friends, which he, Grenville, would present to the King. Morton (a Cambridge man) had already promised to arrange for a testimonial from London, mostly from the Royal Society. Could Shepherd arrange a testimonial, Maskelyne asked, from the appropriate professors in Cambridge?[5]

The original testimonials still survive and make interesting reading. The first, from London undated, signed by a Commissioner of Admiralty (Captain Richard, Viscount Howe, shortly to become Treasurer of the Navy), the President of the Royal Society, with a Vice President, two Secretaries and twelve other Fellows (graduates of Edinburgh, Leyden and Cambridge, but none from Oxford), reads as follows:

The Reverend Mr Nevil Maskelyne Fellow of the Royal Society having approved (*sic*) himself on several Occasions to be well skill'd in the Theory and Science of Astronomy, as a Mathematician, and also vers'd in the practical Part thereof, as a diligent Observer; we judge him to be well qualified to execute the Office of His Majesty's Astronomer at Greenwich.

Howe	Morton
Jno. Blair	Cha Cavendish
C. Moss	Willoughby
S. Demainbray	Matt. Raper
	Jno: Campbell
	W. Heberden
	Tho. Birch
	Geo: L: Scott
	H. Pemberton
	Jno Canton
	Frans. Wollaston
	P. Murdoch
	Cha: Morton
	James Wilson[6]

The second, from Cambridge, was in Shepherd's hand:

Octr. 27 1764

The Reverend Mr Nevil Maskelyne M.A. F.R.S. and Fellow of Trinity College Cambridge by his known skill in the Theory and Practice of Astronomy and Navigation appears to us to be very properly qualified to perform the duty of His Majesty's Astronomer at Greenwich.

R[oge]r Long, Master of Pembroke Hall Cambridge and
    Lowndes's Professor of Astronomy and Geometry.
Robert Smith, Master of Trinity College, Cambridge [and
    Plumian Professor from 1716 to 1760].
Anthony Shepherd, Plumian Professor of Astronomy and
    Experimental Philosophy at Cambridge.
Edward Waring, Lucasian Professor of Mathematicks at Cambridge.[7]

Meanwhile, the President and Council of the Royal Society had been considering a matter raised by Maskelyne himself in June 1763 – the carrying away from the observatory of Bradley's observation books by his executors. Queen Anne's warrant of 12 December 1710 had appointed the Royal Society 'constant visitors of our Royal Observatory' and had authorized them to demand copies of the observations annually. After Flamsteed died in 1719, however, this had not happened and it was the executors' contention that, legally, Queen Anne's warrant no longer applied after her death (an opinion shared by the government's law officers) and that the observation books were therefore the astronomers' private property. So far the Visitors had failed to achieve anything as far as the recovery of Bradley's papers was concerned – and it was to be another thirty years before they did so – but they were determined that such a situation should not be allowed to arise in the future. On 8 November 1764, therefore, the Royal Society submitted a long memorial to the King, detailing the history of the relations between the Monarch, the Royal Society, and the Astronomer Royal since 1710, and proposing a set of Regulations for the new incumbent, summarized thus:

1. The Astronomer Royal (AR) must reside in the observatory and must not accept any additional post which would require him to be absent for any length of time. Absences of more than a few days to be authorized by the Visitors. (Bliss in particular had been criticized for spending so much time in Oxford, where he was Savilian Professor.)
2. The AR and his Assistant never to be absent at the same time.
3. No person to be admitted to the observing rooms unless the AR or Assistant is present.

4. No money to be taken from persons admitted to see the house or instruments.
5. The AR to keep observation books and to authenticate by signing at the bottom of every page.
6. The Assistant to keep and authenticate separate books.
7. Original observation books never to be removed from the observatory under any circumstances. (All previous books had been taken away.)
8. Fair copies of observation books to be delivered to the Council of the Royal Society within six months of the end of every year. (Flamsteed was the only one to have done this previously.)
9. The AR and his Assistant to follow the directions of the Visitors, whether appointed now or in the future.[8]

So, early in November 1764, Maskelyne's testimonials and the Society's memorial reached the Secretary of State's office.

In fact, there was considerable public interest in the vacant post, stimulated no doubt by the ongoing story of Harrison and the longitude awards which received a certain amount of attention in the Press. It seems there were more candidates than just Maskelyne and Betts, as is evident from a poem published in the December 1764 issue of the *Gentleman's Magazine*,[9] the first verse of which reveals that there were originally some ten candidates but that they are now reduced to four:

GREENWICH HOY!
or the ASTRONOMICAL RACERS

> Two lunar months are past, and more,
> Since of these heroes half a score
> Set out to try their strength and skill,
> And fairly start for *Flamsteed-Hill*,
> But lo, from doubts, or fears, or surfeit,
> Six have drawn stakes, or else paid forfeit;
> and thus there now remains no more
> To run the match, than doughty four

One would like to know the names of the six who had drawn stakes!

The next three verses dispose of the chances of Joseph Betts, 46, graduate of Oxford; Nevil Maskelyne, 32, of Cambridge; and James Short, 54, of Edinburgh:

> The first who vaunts the race he gets,
> Is affluent professor B--ts;

> Whose first of *April's* lunar map
> Has giv'n his judgement such a rap,
> As to induce his warmest friend
> To wish no longer he'd contend;
> Who owns the *place* his only view;
> The business journeymen may do.
>
>   The N---b's brother next advances,
> Who, with some mettle, skips and prances:
> But take care, Rev. M-sk-l-n,
> Thou scientific harlequin,
> Nor think, by jockeying, to win:
> Why, when the foremost in the course,
> Woulds't thou thy hopeful chance reverse,
> Avouching with ungen'rous mind
> The two most worthy had declin'd?
> Believe me, this fallacious boast
> Has run thee the wrong side o' the post:
> For the *great donor* of the prize
> Is just, as *Jove* who rules the skies.
>
>   The next, who promised some sport,
> Is the renown'd optician Short;
> Who, cautious, acting like a man,
> Makes all the interest he can,
> And candid hopes, if he should fail,
> Experienced *Nestor* may prevail.

C.P. then reveals the identity of his Nestor – the 71-year-old Oxford graduate John Bevis:

> *Nestor*, aloud, the standers-by,
> Looking around with pleasure cry –
> And will thou, *Bevis*, wilt thou venture
> Against such hardy weights to enter?
>   Yes, clear the course, and call the grooms,
> For lo! how he attended comes

The next ten lines recall his friendship with Newton, Halley and Bradley; and then comes the peroration:

> Tho' no professorship you hold,
> No fellowship endow'd with gold,
> No pension on this worldly stage,
> To comfort thy advancing age,
> Yet has the *Prussia* hero deign'd

> To fix thee 'midst his learned land.[10]
> Courage! then Sir, nor drop thy spirit
> Thy royal master's heard thy merit;
> And that the world with outstretcht eyes,
> Looks on, and points thee for the prize.
> Nay, singly ask the other three,
> On whom (himself excepted) he
> Thinks that the dubious lot should fall,
> *Bevis*, they'll answer – one and all,
> Keep then this adage old in view
> That what all say must sure be true:
> And 'gainst the field, I think we may
> Venture some odds – you get the day.
> C.P.
> *Pall Mall,*
> *December 24th 1764*

Notwithstanding this propaganda, it was Nevil Maskelyne, the scientific harlequin, that his royal master chose from among the doughty four.

The first official news of the appointment of the new Astronomer Royal seems to have been announced at a meeting of the Board of Longitude on 19 January 1765, minuted as follows:

A Letter from the Earl of Sandwich to the Earl of Egmont [First Lord, in the Chair] was then read, setting forth that he had receiv'd His Majesty's Command for the Appointment of the Revd. Mr Nevil Maskelyne to the Royal Professorship at Greenwich, but that he would not be in Possession of his Office till the middle of the next week as the necessary forms of Appointment will take some days before they can pass; that both he and Mr Professor Shepherd who has some papers to lay before the Commissioners, are obliged to be at Cambridge this day on Business of Importance: And that the other Professorship which is now vacant [presumably Bliss's Savilian Chair of Geometry] will probably be filled up soon.[11]

Bureaucracy having taken its course, Maskelyne eventually received his Royal Warrant of Appointment, dated 8 February 1765, the basic wording being the same as in the Warrant addressed to John Flamsteed by King Charles II in 1675:

[signed and sealed] George R.

> George the Third, by the Grace of God, King of Great Britain, France and Ireland, Defender of the Faith &c, To Our Trusty & Wellbeloved Nevil Maskelyne, Clerk, Master of Arts, Greeting:
> We being well satisfied of your Learning, your Industry, and great Skill and Ability in the Science of Astronomy, do, by these Presents, constitute and appoint you Our Astronomical Observator in Our Observatory at Greenwich,

in the Room and Stead of Nathaniel Bliss, deceased, requiring you forthwith to apply yourself with the most exact Care and Diligence to the rectifying the Tables of the Motions of the Heavens, and the Places of the fixed Stars, in order to find out the so much desired Longitude at Sea, for perfecting the Art of Navigation;

And it is Our Will and Pleasure that you forthwith take Possession of Our said Observatory with all Buildings, Yards, Inclosures, & also all Instruments & Utensils and all other Things whatsoever belonging thereunto;

And that you receive the like yearly Salary of One Hundred Pounds as was received by the said Nathaniel Bliss, the same to be paid unto you by Our Treasurer of Our Ordnance by four equal Quarterly Payments at the Four most usual Feasts, The first Payment to be made at Lady Day next ensuing the Date hereof, and so to continue during Our Pleasure, & that further for your Assistance in the Execution of the laborious Part of your said Office, you likewise be allowed and paid in the same manner the yearly Salary of Twenty Six Pounds for such Servant or Labourer whom you shall make use of for that purpose in like manner as was allowed, & paid to, or for, the Servant of the said Nathaniel Bliss, It being Our Royal Intention that you shall have, receive, and be allowed all and every such Salaries, Payments, Advantages and Privileges, as were at any time, had, enjoyed, or received by, or allowed to the said Nathaniel Bliss, as Our Astronomical Observator, or as Astronomical Observator of any of Our Royal Predecessors.

Given at Our Court at St James's the Eighth day of February 1765 In the Fifth Year of Our Reign.

By His Majesty's Command.

[signed] Sandwich

[To] Nevil Maskelyne, Astronomical Observator.[12]

So, on Friday, 8 February 1765, at the age of 32, Nevil Maskelyne became England's fifth Astronomer Royal.

The very next day, Saturday 9 February, he was at the Admiralty in London, attending his first meeting of the Board of Longitude in his official capacity. This was probably the most important meeting of that body in the 114 years of its existence, because awards were recommended for Harrison, Mayer and Euler, and Maskelyne presented his proposals for the publication of a Nautical Almanac and Astronomical Ephemeris, thereby starting the chain of events which led to the Greenwich Meridian becoming Prime Meridian of the World and the basis of the world's system of timekeeping.

The decisions taken at that important meeting – which led to a new Longitude Act effectively changing the rules in the middle of the game – will be recounted in Chapter 8. Meanwhile, however, we should conclude this chapter by detailing the other formalities connected with the change of Astronomer Royal. On 13 February, the new Astronomer Royal waited upon the King and kissed hands upon appointment.[13] On 22 February, the King signed a Royal Warrant appointing the President and Council of the Royal Society as constant Visitors of the Royal Observatory, thereby renewing the lapsed Warrant of Queen Anne[14]. Then, on 5 March, Sandwich wrote to the Marquess of Granby, Master General of the Ordnance, informing him of Maskelyne's appointment and ordering the Board of Ordnance to pay the salaries and repair or replace the instruments as asked for by the Visitors.[15]

On 16 March, the King signed a Warrant entitled 'Regulations for the Astronomical Observator at Greenwich',[16] based on the memorial submitted by the Royal Society described on p. 55 above, with the additional proviso that, if the Astronomer Royal wished to absent himself from the Observatory for more than ten days, this had to be agreed in advance with the Secretary of State as well as the Visitors. Before the Warrant was signed, Maskelyne was shown the whole text and was not best pleased. He wrote to the Royal Society on 28 February, hoping that they would not propose further regulations as this could undermine his authority.[17] Then, to add insult to injury, when he went to the Secretary of State's office to collect the Warrant, he was asked to pay fees of £8 1s. 6d. He sent the bill to the Society. A year later, they were still quibbling and Maskelyne had to remind the Society that the new regulations were 'not made at my desire but at your suit.[18] He got his money eventually.

Charles Green, Maskelyne's companion in Barbados, had returned to Greenwich to continue his post as Assistant to the Astronomer Royal. While he had been away (from September 1763 to July 1764), someone had continued to make the regular observations. Whom this was, we do not know. It could not have been Bliss himself because of his declining health, but possibles, all of whom had had suitable experience (though they would have had to take time off from their regular employment) would be Gael Morris, John Bradley, John Robison or Robert Waddington. The handwriting in the surviving observation book is of no help as it seems to be a copy of the original.[19] Anyhow, Green continued the observations at Greenwich until 9 March 1765.

Maskelyne took up residence officially in the observatory on 16 March and the same day there was a Visitation by the Royal Society,[20] attended by both Maskelyne and Green. On Lady Day, 25 March, Green left and Joseph Dymond joined as Maskelyne's Assistant – the servant or labourer of the Warrant. As will be described later, the first observations during Maskelyne's term of office were not made until 7 May 1765.

When the post of Astronomer Royal had been established in 1675, the salary was set at £100 a year, plus £26 for the servant or labourer, paid by the Board of Ordnance. This was the only emolument from the post enjoyed by John Flamsteed, first Astronomer Royal, who was also expected to provide his own instruments, though he was lucky enough to find a patron, Sir Jonas Moore, Surveyor General of the Ordnance, to do this for him initially. In 1729, as a result of a visit by Queen Caroline, the second incumbent, Edmond Halley, was granted in addition half pay as a naval Captain (payable from Navy votes), in which rank he had served from 1698 to 1701 when in command of the *Paramore* pink: this pension died with him. On James Bradley's appointment in 1742, he only had the £100 salary from the Ordnance, though, like Halley before him, he was also receiving emoluments from the post of Savilian Professor at Oxford. Then, from Christmas 1752, King George II granted him an additional pension of £250 a year (payable at the Exchequer), which was continued for his successors Bliss and Maskelyne.[21] The latter's total gross emoluments for the post were thus £350 a year, reduced by fees and taxes to a net sum of about £307 a year.[22] Out of this, he had to pay for the Assistant's board and keep.

In January 1765, the Royal Society and Maskelyne himself tried hard to get the Treasury to agree to his assistant's post being officially upgraded from 'servant or labourer' to 'assistant astronomer' in line with the actual duties performed, so that the derisory £26 per year might be increased.[23] The Treasury refused to sanction any increase of salary at that time and the title was not changed in the appointment warrant, though in the Regulations, he was called Assistant. The Treasury eventually relented, raising the Assistant's salary to £70 p.a. in 1771 and £100 in 1810. The Astronomer Royal's emoluments remained the same until after Maskelyne's death.

7

The Royal Observatory and its instruments

In the autumn of 1674, 'an accident happened that hastened, if it did not occasion, the building of the Observatory.' So wrote John Flamsteed, first Astronomer Royal, many years later, continuing: 'A Frenchman, that called himself Le Sieur de St Pierre, having some small skill in Astronomy, and [having] an interest with a French lady, then in favour at Court, proposed no less than the discovery of the Longitude.'[1]

The lady in question was none other than the 25-year-old Louise de Keroualle, a Breton lady who was King Charles II's mistress most in favour at that time, whom he had created Duchess of Portsmouth the previous year. The King asked a Royal Commission to investigate St Pierre's claims: they reported that the method (which involved measurements of the altitude of the Moon and certain stars), though not unsound in theory, could not be used in practice because neither the positions of the stars nor the theory of the Moon's motion against the background of the stars were then known to an adequate precision. But, they said, if the King were to found an observatory (as King Louis XIV of France had done some eight years earlier), then, in the fullness of time, the necessary data could be obtained and finding longitude at sea would become possible. 'When Charles II, King of England, was informed of these matters', wrote Flamsteed on another occasion, 'he said the work must be carried out in Royal fashion. He certainly did not want his ship-owners and sailors to be deprived of any help the Heavens could supply, whereby Navigation could be made safer.'[2]

On 4 March 1675, the 27-year-old John Flamsteed was appointed the King's 'astronomical observator' and poor St Pierre was sent packing without the hoped-for reward. So a King's mistress and an unknown Frenchman – his identity is still a mystery – acted as unwitting catalysts in the foundation of England's national observatory and first government scientific establishment.

On 22 June, the King addressed a warrant to Sir Thomas Chicheley, Master General of the Ordnance:

Whereas, in order to the finding out of the Longitude of Places for perfecting Navigation and Astronomy, we have resolved to build a small Observatory in Our Park at Greenwich upon the highest Ground at or near the Place where the Castle stood, with lodging rooms for our Astronomical Observator and Assistant. Our Will and Pleasure is that according to such Plot and Design as shall be given you by Our Trusty and Well-beloved Sir Christopher Wren Knight, Our Surveyor General, of the Place and Scite of the said Observatory, you cause the same to be fenced in, built and finished with all convenient speed . . . and that you give order unto Our Treasurer of the Ordnance for the paying of such Materials and Workmen as shall be used and employed therein, out of such Monies as shall come to your Hands for old and decayed Powder, which hath or shall be sold . . . provided that the whole sum . . . shall not exceed five hundred Pounds . . .'[3]

So was built – 'with bricks from Tilbury Fort where there was a spare stock; and wood, iron and lead from a gatehouse demolished at the Tower',[4] and with money raised from the sale of 690 barrels of old and decayed gunpowder at Portsmouth – the building we know today as Flamsteed House, in the words of the designer Christopher Wren, 'for the observator's habitation, and a little for pompe',[5] destined to become Nevil Maskelyne's official residence. The full story of the foundation and early years of the Royal Observatory is told elsewhere.[6] Suffice it to say here that, on the first floor of Flamsteed House, Wren designed the elegant octagonal Great Room, used for observing with portable instruments, which were moved from window to window as necessary. The bedrooms and living rooms were on the ground floor, the kitchen and cellars in the basement, while the serious observational work took place in the Sextant and Quadrant Houses, small adjoining buildings at the bottom of the garden.

Flamsteed started regular observations – for settling the places of the stars, and improving the Theory of the Moon's Motion – in October 1676, and continued until his death on the last day of 1719, his observational results and the star catalogue resulting from them (used actively by astronomers for well over a hundred years) being published posthumously in 1725.[7] At this point, it must be emphasized that the observations made in observatories involve, not just gazing through telescopes, but making measurements – at that date, measurements of time and measurements of angular distances in the heavens.

When Edmond Halley came to Greenwich after Flamsteed's death, he found it devoid of instruments, the latter's widow having removed all of them on the grounds that they were private property. Halley obtained a government grant of £500 to re-equip the observatory. From the cele-

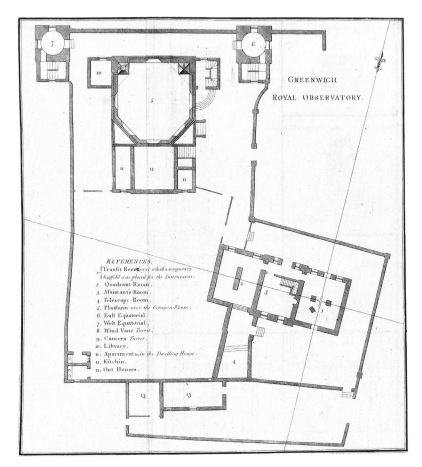

Fig. 7.1. Plan of the Royal Observatory, 1788. Flamsteed House is at top centre, Bradley's New Observatory at bottom right. Taken from General Roy's report of the triangulation carried out in 1787 to connect the observatories of Greenwich and Paris, when he erected his great theodolite precisely over the transit instrument. This plan shows the state of the observatory buildings generally from about 1750 to 1808, when the Circle Room was built onto the east end of the Transit Room. From *Phil. Trans. R. Soc.*, **80**, plate 12. (Photo: NMM)

brated clock and instrument maker George Graham, he obtained in 1725 a *mural quadrant* of 8 feet radius and three clocks, the former mounted on a stone meridian wall which he caused to be built, all of which survive to this day. However, £500 was not enough to provide all that was necessary, so Bradley, who succeeded Halley at Greenwich in 1742, obtained another grant in 1749, this time of £1000 'to be paid by

Fig. 7.2. *The Royal Observatory from the South East*, c.1765. Though this unsigned and undated oil painting must have been executed some years before the plan in Fig. 7.1, few changes occurred in the interval. The viewpoint is from bottom right of Fig. 7.1, whose reference numbers are used below.

The right-hand building is Bradley's New Observatory showing (left to right) the Telescope Room (4) with Quadrant Room (2) behind, Assistant's Rooms (3) with bedroom on the top floor, and Transit Room (1).

Flamsteed House is the building with the tall chimney behind, showing the turrets for the Wind Vane (8) and Camera Obscura (9) on the roof. The square building on the left, with a hemispherical roof, seems to be the western summer house before its 1773 conversion to take the West Equatorial telescope (7), but painted to be greatly exaggerated in size.

From an oil painting in possession of the Royal Greenwich Observatory. (Photo: RGO)

the Treasurer of the Navy out of money arising from the old stores of the Navy'.[8] Shades of old and decayed powder! With this grant, he obtained a second mural quadrant, an 8-foot *transit instrument*, and a *movable quadrant* of 40 inches radius, all from John Bird, a 6 foot focus reflecting telescope from James Short, and a *transit clock* made under Graham's supervision (and with his name on the dial) by John Shelton.

Bradley also persuaded the Board of Ordnance (which, for historical reasons until 1819, was responsible for the maintenance of the buildings and instruments of the Royal Observatory) to build the 'New Observatory' to accommodate the fundamental instruments and provide accommodation for the Assistant, aligned to the meridian (unlike Flamsteed House) south of the present courtyard. There were three rooms at ground level: at the west end, the Quadrant Room, built over Halley's meridian wall; at the east end, the Transit Room; and in the centre, the Assistant's library and calculating room with his bedroom above. The New Observatory became operational in 1750. In Bliss's time, the building of a small observatory with a revolving dome for the movable quadrant was started south of the Quadrant Room on the site of Flamsteed's Sextant House.[9]

So, when Maskelyne took up residence in March 1765, he inherited three buildings: Flamsteed House, with three, perhaps four bedrooms; the New Observatory described above; and the just-completed (though not yet equipped) movable quadrant observatory (Figs. 7.1 & 7.2).

Before describing the instruments he inherited, we must consider the observations that had to be made, the most important ones being those needed to define the positions of bodies in the heavens. On the surface of the Earth, positions are defined by two coordinates: terrestrial latitude, 90° north or south of the equator; and terrestrial longitude 90° east or west of the Greenwich meridian. In the heavens, the equivalent equatorial coordinates are *declination*, 90° north or south of the celestial equator, and *right ascension*, 24 hours measured eastwards (generally in units of time – hours and minutes rather than degrees and arc-minutes) along the celestial equator from the *First Point of Aries*. (There had been a system of ecliptic coordinates using celestial latitude and longitude, but, for astronomers, this had fallen out of use by this date.) Both these coordinates could be obtained when a body was observed crossing the meridian, exactly north or south:

(a) declination by measuring its *zenith distance* with a mural quadrant (Fig. 7.4);
(b) right ascension by timing the precise moment it crosses the meridian, as seen in a transit instrument (Fig. 7.5).

Such observations constitute what is today known as fundamental – or positional – astronomy, which was the main interest of astronomers until the advent of photography and spectroscopy in the nineteenth century made possible the study of astrophysics – what heavenly bodies were made of, their life, and their death.

As well as barometer and thermometer readings both inside and out-

Fig. 7.3. *The Old Royal Observatory Courtyard*, 1982. The Greenwich meridian today passes through the tall doors of Airy's Transit Circle Room beneath the smaller dome. The building to the right of that is Bradley's New Observatory, now restored to the state it was in Maskelyne's time, showing (from left to right) the Transit Room, the Assistant's calculating room and library with bedroom above, and Quadrant Room. From a watercolour by Terence Scales, dated 1982, in possession of the author. (Photo: NMM)

side the observing rooms, two other types of observation were made at Greenwich in Maskelyne's time:

(c) the timing of lunar and solar eclipses, the eclipses of Jupiter's satellites, occultations of stars and planets by the Moon, and the transits of Venus and Mercury, as seen through simple 'gazing' telescopes, generally in the Great Room (later in the Advanced Room, to be described), all used for finding the longitude difference between Greenwich and anyone else who made corresponding observations of the same phenomena at the same time; and

(d) measuring the positions of comets relative to nearby stars whose own positions are known, using an *equatorial sector*, also in the Great Room.

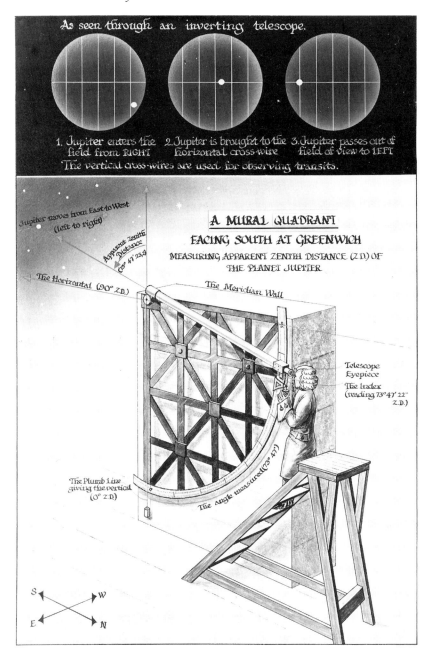

Fig. 7.4. Measuring Zenith Distance with the Mural Quadrant. (Photo: NMM)

The Royal Observatory and its instruments 69

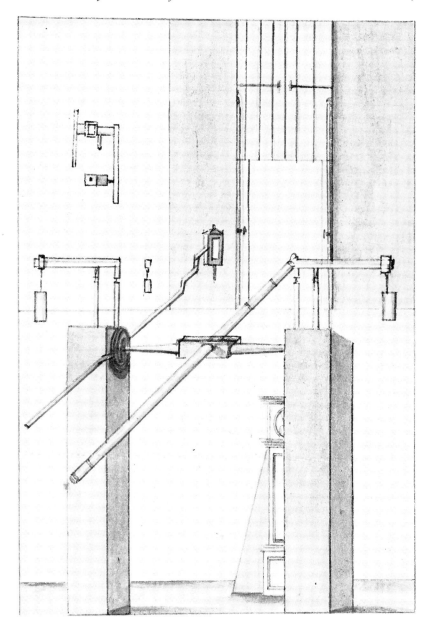

Fig. 7.5. *The Transit Room*, c. 1785, showing the 5-foot transit instrument by John Bird and clock by George Graham. From a pen-and-wash drawing by the naval author John Charnock of Blackheath, preserved at the National Maritime Museum (Charnock's Views, vol. 4). (Photo: NMM)

The principal instruments for which Maskelyne became responsible were the transit instrument and the two mural quadrants in Bradley's New Observatory. A complete list, from the inventory taken at the Maskelyne's first Visitation on 16 March 1765, is given in Appendix D.

Maskelyne found the observatory in a very run-down state, neither Bradley in his later years nor his successor Bliss having given the buildings and instruments the attention they deserved; furthermore almost all the observing had been left to the Assistants. A fortnight after the Visitation, on 28 March 1765, Maskelyne sent to the Royal Society for forwarding to the Board of Ordnance a list of the essential cleaning and repairs needed and asked additionally for two journeyman clocks.[10] The Society seems to have taken no immediate action, which caused Maskelyne to write a stiff letter on 13 May saying that he could not take any observations until the Board of Ordnance (to whom the Society ought to have sent Maskelyne's requirements) did something, adding: 'I shall propose to apply to higher authority for this purpose, in which I hope for your assistance.'[11] Thus stimulated, the Society's Council met on 16 May and drafted a letter and list, which the Astronomer Royal himself took to the Office of Ordnance. The Board immediately said that they had never before dealt with instruments and anyhow there was no money in the budget for that purpose. Four long months later, following much behind-the-scenes lobbying with the Government, the Board reluctantly agreed to bear the necessary expenses.

Despite these difficulties, observations had in fact started on 7 May and Bird seems to have done a certain amount of cleaning and repairs before the Board of Ordnance accepted responsibility.

On 25 September 1765, there occurred in the Great Room of the Royal Observatory an event not unimportant in the history of astronomy – the observation of the immersion of the star Delta Capricorni into the Moon's dark limb by 'a 3½-foot telescope of Dollond with 3 object glasses'.[12] This was the first recorded observation at Greenwich by an *achromatic telescope* (invented – though there is some dispute on this score – by John Dollond in 1758), taken with the first *triple-achromatic* ever made by John's son Peter[13] to a design which was to prove so successful both technically and commercially. (Fig. 7.6)

In his list of wants, Maskelyne had suggested that the missing ten-foot object glass should be replaced by 'one of Mr Dollond's new combined object glasses'.[14] This was not immediately forthcoming and the 3½ft. telescope seems to have been provided instead. There is no record of

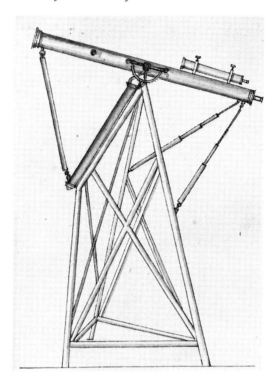

Fig. 7.6. *The 46-Inch Triple-Achromatic Telescope by Peter Dollond*, c. 1785. Another drawing by John Charnock. (See Fig. 7.5). (Photo: NMM)

payment for this telescope, so one presumes that Peter Dollond either presented or lent it to Maskelyne, so that he, Maskelyne, could give this prototype a thorough trial.

Dollond's investment paid off. His telescope proved so successful that hundreds were sold all over Europe (and a few in America) over the next few years. Almost overnight, all non-achromatic refractors (some up to 25 feet long, and including those on large quadrants and transit instruments in observatories) became obsolete, to be replaced by the shorter and lighter achromatic instruments.

Maskelyne's observing policy differed somewhat from that of his predecessors. Highest priority was given to the Moon, the Astronomer Royal and his Assistant taking observations, the former with the transit instrument, the latter with the mural quadrant, on every possible occasion that she crossed the meridian. When only one observer was available, priority was generally given to the transit instrument. The Sun and planets were observed likewise, though at somewhat lower priority. As for the stars, both Flamsteed and Bradley had observed places of more than 3000 individual ones: Maskelyne decided that enough such data existed on these for the time being and, except in special circumstances, limited his observations of the stars to 31 (later increased to 36) very special ones needed to ascertain the going of the clocks. These were chosen (a) to lie near the celestial equator so that their apparent *diurnal movement* was as great as possible, and (b) to be sufficiently bright to be observable in a telescope in daylight, giving the greatest possible chance that at least one of these 'clock stars' would be visible every 24 hours.

One of the results of this policy was that observing after 10 p.m. was generally restricted to those nights when the Moon crossed the meridian after that time – only a few nights each month, and then only if they were clear. On these occasions, the opportunity was taken to observe stars and planets as well, an hour or so either side of the lunar transit.

It should be emphasized that the actual observing is only the start of the process. The results then have to be 'reduced', by having corrections applied, and copied into fair register, perhaps having the resulting right ascensions and declinations computed, etc. – all of which kept the Assistant busy when he was not actually observing. To give some idea of the observing work load, there follows a summary of three typical days' worth of observations:

(a) 1783 March 22 – Sun at noon, stars at 4.23 p.m., 5.04, 5.12, 5.43, 6.35, 7.28, 7.32, 9.17, 9.57, 3.25 a.m., 3.33, Moon at 3.38 a.m., star at 7.40. Mural quadrant observations of Sun and Moon only.[15]

(b) 1797 March 21 – Sun at noon, stars at 4.20 p.m., 4.58, 5.40, 7.24, 7.29, 6.26 a.m., Moon at 7.37, star at 8.30. Quadrant observations of all, plus four more bright stars.[16]

(c) 1797 Sept. 21 – Sun at 11.54 a.m., Spica at 1.14 p.m., Mercury at 1.27, Arcturus at 2.06, Alpha Capricornus at 8.06. Quadrant observations of Sun and Mercury only.[17]

The numbers and frequency of observations in the Great Room (later in the Advanced Room) were very variable. In 1771, for example, there were five observations of occultations and 14 of Jupiter's satellites; there were no comets, the previous one being in 1769, the next in 1773. In

1802, there were four occultation observations (including one of five stars of the Pleiades), 11 of Jupiter's satellites, one lunar eclipse, and one transit of Mercury; there were ten observations of the newly-discovered minor planets Ceres and Pallas with the equatorial sector; once again, no comets, the previous one being 1799, the next 1807.[18]

We will conclude this chapter by reprinting from a London guide book, a description written by Thomas Evans, who was Assistant from 1796 to 1798:

The Observatory is composed principally of two separate buildings, one of which is the Observatory properly so called, where only the assistant lives, and makes all the observations; the other is the dwelling-house, in which the astronomer-royal himself resides. The former being the most essential, we shall describe it first. It consists of three rooms on the ground floor, the middle one of which is the assistants' sitting and calculating room, furnished with a small library of such books only as are necessary for his computations, and a clock made by the celebrated Graham, which once served our immortal Halley as a transit clock. The face, which resembles one described by Ferguson, is the only curious part of it.

Immediately over this is the assistant's bedroom, with an alarm to awake him to make his observations at the proper time. Nothing can exceed the tediousness and *ennui* of the life the assistant leads in this place, excluded from all society, except, perhaps, that of a poor mouse which may occasionally sally forth from a hole in the wall, to seek after crumbs of bread, dropt by his lonely companion at his last meal! This, of course, must tend very much to impede his acquiring astronomical information, and damp his ardour for those researches which conversation with scientific men never fails to inspire. Here forlorn, he spends days, weeks, and months, in the same long wearisome computations, without a friend to shorten the tedious hours, or a soul with whom he can converse. He is also up frequently three or four times in the night (an hour or two each time,) and always one week in the month when the moon souths in the night time, with the owls perched on the fir-trees in the park below, screaming by way of answer to him when he opens the sliding shutters, in the roof of the building, to make his observations. A zealous wish on *his* part to promote so divine a science as that of astronomy, joined to an awful contemplation of the wonderful works of the Almighty, are the sole objects that afford him pleasure in this solitary hermitage. . .[19]

Poor, lonely Thomas! However, we need not be too sorry for him. It was not long before he did find a friend, falling in love with and marrying Margaret Maskelyne's 23-year-old governess, Deborah Mascall, while still at the observatory – but that is another story.[20]

8

The Harrison affair, 1764–7

We will now take up again what had become quite a drama – the Harrison affair. It will be recalled that William Harrison arrived back in England with H4 on 16 July 1764, while Maskelyne had remained in Barbados (Chapter 5). During the whole voyage to Barbados and back, the watch had behaved as well as or better than the Harrisons could have hoped and they were in an optimistic mood when William waited in the ante-room at the Admiralty before the meeting of the Board of Longitude on 18 September. And the results of the meeting seemed to bear out their optimism, as witness this letter written two days later by William Harrison to his father-in-law, Robert Atkinson:

They (the Board) were all as agreeable as could be, Parsons and all, as they have now lost their ringleader; they gave us a thousand pounds as by agreement, and then nominated three Gentlemen on their part to compute the Observations, and ordered me to nominate three on our part, which six Gentlemen are to give their computations on or before the last day of October; then another Board will be called.

The Chair is dead; as to the Moon, there is nothing said about it. I have to thank God as fine a Prospect before me as can. . .[1]

The Parsons were, of course, the Professors on the Board, the ringleader being Bliss who had died on 2 September. The thousand pounds was the second part of the £2500 awarded as laid down in the Act on 17 August, to be given after the second sea trial. To do the computations to find the difference of longitude between the Portsmouth Royal Academy and Maskelyne's observatory in Barbados, using the corresponding observations of occultations and of the eclipses of Jupiter's satellites taken by John Bradley at Portsmouth and Maskelyne and Green in Barbados (taking into account corresponding observations at Greenwich), the Board nominated Captain John Campbell RN, Dr John Bevis and George Witchell of Portsmouth. Harrison had only one nominee, James Short.

As for the Chair, Christopher Irwin's son attended the same meeting, complaining of his father's ill usage in the *Princess Louisa*, asking for a second trial. Asked what his complaint was, he said that Maskelyne had made no observations during the voyage and had prevented Green from making more than two. However, the Board dismissed him, saying they did not think 'there was the least possibility of his Chair answering the purpose for which it was designed' and therefore recommended him 'to desist from pursuing the said Invention any further.'[2]

A month later, Maskelyne arrived back in England, full of enthusiasm for the success of his lunar observations – to hear that Bliss was dead.

The next meeting of the Board of Longitude was on 19 January 1765 and the first business was the announcement of Maskelyne's appointment as Astronomer Royal. Then the Board went on to consider the results obtained by the four people appointed at the last meeting to do the computations to determine H4's performance. Their results were almost unanimous, the mean value for the difference of longitude between Portsmouth and Barbados being 58°34'.5 by astronomical observations[3], 58°44'.4 by the watch, a difference of 9'.9, equivalent to 8.5 geographical miles, well within the limit set by the Act of Queen Anne for the largest award (30 geographical miles). After hearing a Memorial from John Harrison, the First Lord adjourned the meeting for three weeks, asking that all Commissioners should do everything in their power to attend this next meeting.

The Harrisons did not attend that meeting but the news of the new appointment soon reached them. The new Astronomer Royal was not to be Bevis, or Short – both good friends to the Harrisons – but, of all people, Maskelyne – not just another parson, but someone whom they knew favoured the rival longitude method, and might actually be competing. Their optimism of last September evaporated.

At a Meeting of the Commissioners appointed by Act of Parliament for the discovery of the Longitude at Sea &c which was held at the Admiralty on Saturday the 9th of February 1765.

Present

Rt. Honble Earl of Egmont	First Lord Commr. of the Admty.
———— Sr. John Cust Bt	Speaker of the Ho. of Commons
———— Ld. Visct. Barrington ..	Treasurer of the Navy
Sir William Rowley KB	Admiral of the Fleet

Henry Osborn Esqr ...	Admiral of the White
Charles Knowles Esq.	⎫
Honble John Forbes	⎬ Admirals of the Blue
Sir George Pocock KB	⎭
Rt. Honble Earl of Morton	President of the RI. Society
Revd. Mr. Maskelyne	Royal Astronomer
Revd. Mr. Hornsby	Savilian Professor of Astronomy at Oxford
Mr. Waring	Lucasian Professor of Mathematicks at Cambridge
Revd. Mr. Shepherd	Plumian Professor of Astronomy Experiml. Philosophy at Do.
Philip Stephens Esqr. ...	Secretary of the Admiralty
George Cockburn Esq .	Comptroller of the Navy[4]

That was how the minutes of the momentous meeting of the Board of Longitude on 9 February 1765 were prefaced. A distinguished gathering indeed – all those who had attended three weeks earlier, plus Shepherd from Cambridge, and the newly-appointed Astronomer Royal, Maskelyne.

After routine business, Harrison's memorial from the last meeting was read again. He claimed that the terms for the highest award in the Act of Queen Anne had now been complied with and asked that the Board sign a certificate to that effect so that he could collect his award from Parliament – £20 000, less the £2500 already granted to him. The Board agreed unanimously that 'the said Timekeeper has kept its time with sufficient Correctness & without losing its Longitude in the Voyage from Portsmouth to Barbadoes beyond the nearest Limit required by the Act of the 12th of Queen Anne, but even considerably within the same'. However, they pointed out that that same Act also stipulated that the method should be 'found Practicable and Useful at Sea'. They could not possibly sign a certificate to that effect on the evidence of the performance of a single timekeeper on a single occasion. They must be certain that other workmen could make timekeepers to Harrison's principles with a similar performance (and at a reasonable cost). For them to be certain, Harrison must make known these principles, which, as Lord Morton pointed out, he still had not done although required to do so (originally at his own suggestion) by the Act of 1763.

What they *would* do, the Board said, was to recommend to Parliament the award of £10 000 (minus, of course, the £2500) 'upon his producing his Timekeeper to certain persons to be named by the Board & discovering to them, upon oath, the principles & manner of making the same'; and that the other £10 000 should be awarded 'on proof [unspecified] ... that his method will be of common & general Utility in finding the Longitude at Sea...'[5]

Having informed the Harrisons of these decisions, the Board then went on to consider a long Memorial from Maskelyne praising the efficacy of the lunar method using Mayer's last manuscript tables, as he himself had proved at sea. Four officers from East India Company ships were then examined on the utility and practicability of the method as described in the *British Mariner's Guide*, in which they all had practical experience at sea. They gave evidence on the accuracy of the results achieved, how they could 'make the Calculations in a few hours not exceeding four hours', and saying that they were 'of the opinion that, if a Nautical Ephemeris was published, this Method might be easily & generally practiced by Seamen.' The Board accordingly resolved to recommend to Parliament, (*a*) that Mayer's last manuscript tables should be published and that his widow should be awarded a sum not exceeding £5000, and (*b*) that the Board should be given 'power to give a Reward to persons to compile a Nautical Ephemeris and for Authority to print the same, when compiled, in order to make the said Lunar tables of General Utility.'[6]

As Forbes has pointed out[7], the Board's assessment at that time is indicated by its resolution to recommend awards not exceeding £10 000 and £5000 to Harrison and Mayer respectively. Harrison's invention had made the chronometer method (as it came to be called later) practicable but not generally useful, whereas Mayer's tables made the method of lunars generally useful but not practicable, with the accuracy of the latter only about one half that of the former. Hence, it was felt that Harrison was entitled to half the *maximum* award, and Mayer's heirs half the *minimum* award.

But Harrison fiercely opposed the Board's interpretation. He published another pamphlet[8] bringing up to date his *An Account of the Proceedings* ... of 1763. He also presented a Petition to Parliament, dated February 25 1765, claiming he was entitled to the whole reward there and then. Two days later, according to William Harrison, he received a message from Lord Morton: 'Without he will agree to the Resolutions of the Board of Longitude on the 9th Instant that this affair shall never come to a decision in the House but only put off from time to time.'[9]

In passing, it should be noted that, on 1 February, Lord Morton and Nevil Maskelyne were among those who signed a proposal that William Harrison should be elected to the Fellowship of the Royal Society.

Morton did indeed begin his delaying tactics in Parliament. The Harrison affair was due to be debated in the House of Commons on 5 March, and all the witnesses were assembled. When Morton came to the Lower House from the Lords, he said his eyes were too weak to read his papers and could the matter be put off to another day? Certainly, said the House, we will debate it tomorrow. When tomorrow came, Lord Barrington, Treasurer of the Navy (and an MP), arrived with the news that, as Morton had not received an official summons, he was not prepared to come that day, but could the matter be put off yet another day? This time, however, Honourable Members were not so accommodating, several suggesting that quite likely Morton's evidence would not be needed anyway, and if it were, then he could be heard later.

Maskelyne tells us of this in a somewhat puzzled letter to Shepherd in Cambridge[10] – puzzled why Morton did not seem to want to give evidence to the House, and also refused to see him when he called; 'His Lordship returned an answer that he was engaged: indeed the servant told me before that he was not at home. I think he cannot have any reason to be offended at me.' Perhaps it was with the House of Commons that Morton was offended.

That same letter describes at length the hour's questioning Maskelyne received when the House was resolved in Committee – about his own part in the Barbados trials, his lunar observations at sea, and his opinion as to the relative merits of the two methods. New factors which seem to have emerged were:

(a) that Leonhard Euler, the Berlin mathematician, had produced equations on Newton's principles without which Mayer's tables would have been much less accurate;

(b) that Mayer's tables, though excellent as they stood, were still capable of improvement;

(c) that one of the advantages of the lunar method was that an error in today's results was unlikely to affect tomorrow's, whereas errors with a timekeeper were cumulative; and

(d) that the two methods were not mutually exclusive: happy was the navigator who had access to both (a cardinal principle which still applies even in these days of electronic navigational aids).

On 20 March, the Secretary of the Admiralty gave the report of the Committee of the Whole House, which had heard much additional evidence, including that of William Harrison and Captain John Campbell RN.

Finally, the Act 5 Geo. III *c*.20 – the new third Longitude Act – became law on 10 May 1765. Having noted that the Act of Queen Anne, drafted fifty years earlier, was by now somewhat ambiguous, the new Act generally endorsed the Board of Longitude's proposals of 9 February. There were, however, certain important alterations and additions, italicized below:

(a) the first £10 000 (less £2500) should be awarded as soon as Harrison had explained his principles, but this had to be *within six months* of the passing of the Act, and the four time-keepers had to be handed to the Board as the property of the Public at the same time;

(b) the second £10 000 could be awarded so soon as '*other Time Keepers* of the same Kind shall be made, and shall, *upon Trial* found to be of sufficient Correctness [as specified in the Act of Queen Anne]. . .': who should make the timekeepers was not specified, nor, critically for Harrison, were details of the trials required;

(d) a sum of £3000 (not £5000) should be awarded to Mayer's heirs;

(e) £300 should be paid to Leonhard Euler;

(f) up to £5000 should be paid in the future to anyone who could *improve Mayer's tables*;

(g) the Board should publish a Nautical Almanac and money was made available for computing it: any other person publishing such an almanac would be liable to a fine of £20 a copy (something which is still on the Statute Book).

Parliament called this an Act for explaining and rendering more effectual the two previous Acts of 1714 and 1763. The Harrisons called it changing the rules in the middle of the game.

The Board met on 28 May 1765 to decide the procedure for the discovery of the principles of the watch. There followed three more meetings and two months of tedious and acrimonious exchanges between Harrison and the Commissioners before the Board's representatives assembled at his house in Red Lion Square on Wednesday 14 August – on the one hand three Gentlemen skilled in Mechanics, Rev. John Michell, Rev. William Ludlam, John Bird; three watchmakers, Thomas Mudge, William Matthews, Larcum Kendall; and one Commissioner, Nevil Maskelyne: and on the other hand, John and William Harrison. For four long days –

Wednesday, Thursday, Friday, Saturday – John Harrison began to take the timekeeper to pieces, explaining the function of each part and answering questions. This continued on Tuesday 20th and Thursday 22nd, and the Committee wanted to reassemble on Monday 26th. John Harrison, however, had had enough. He pointed out that the Board had said the Committee should go on without interruption until all were satisfied. So please would they come back tomorrow, Friday, not wait till Monday? At that, the Committee members decided they too had had enough too, so, on that very Thursday, 22 August, they signed a certificate – Ludlam rather reluctantly – to the effect that Harrison's demonstrations and explanations were to their satisfaction.[11]

That certificate was laid before the Board on 12 September 1765. Before receiving the Board's own certificate for the balance of the £10 000, Harrison was required by the Act to hand over H4 and the three earlier timekeepers. It was decided that H4 should be placed in the care of Larcum Kendall with whom – at Harrison's own suggestion – they were negotiating to make a copy. As Kendall lived in Furnival's Inn Court, Holborn, close to Red Lion Square, it would be easily accessible to Harrison who had said he was about to start the construction of two more timekeepers. However, their initial negotiations with Kendall broke down. Nevertheless, at their meeting of 28 October, Harrison handed over H4 and the drawings. Having promised to deliver the three earlier timekeepers, Harrison was given a certificate to enable him to receive £7500. H4 was sealed in a box and placed in the custody of Secretary of the Admiralty. The drawings were handed to Maskelyne, who was ordered to prepare them for publication.

During the winter of 1765-6, things remained quiet on the Harrison front for the first time in several years (John was now 73). It was assumed that he was working on the two watches already mentioned (though only one was ever seen), and *The Annual Register* for 1765 reported that the King of Sardinia had commissioned four duplicates of H4 at £1000 each (which were certainly never delivered).[12]

Then, in February 1766, Ferdinand Berthoud came once more to London, bringing with him this time an offer from Choiseul on behalf of the French Government of £500 sterling if Harrison would divulge his secrets. 'For such a bagatelle', Harrison is reported to have said, 'I refuse'.[13] Berthoud did, however, meet Thomas Mudge, with whom he was in correspondence. They dined together at the house of the Count de Brühl, the Saxon Minister in London, who acted as interpreter. Mudge, who had been one of those to receive Harrison's discovery, explained to Berthoud the principles of H4 with the aid of pencil sketches.[14]

When the Board met on 26 April 1766, they were presented with yet another letter from Harrison, this time complaining that H4 was neglected and lying dormant in a box in the Admiralty while he could do with it to help in making his new watches. He also asked for £800 so that he could set up a manufactory for marine watches. Ignoring Harrison's proposals, the Board took two important decisions: first that H4 should be sent to Greenwich for a protracted trial under Maskelyne's supervision; and secondly, at Ludlam's suggestion, that it should be Larcum Kendall who should be commissioned to make a copy of it.

Ten days later, on 5 May, Maskelyne, accompanied by Captain Thomas Baillie of Greenwich Hospital (an establishment for naval pensioners on the riverside below the Royal Observatory); John Ibbetson (Secretary of the Board of Longitude) and Larcum Kendall, came to the Admiralty where Stephens, the Secretary of the Admiralty, handed over H4 in its box. It was carried from the Admiralty to Whitehall Stairs (about 600 yards) and embarked in the barge of Admiral Sir George Rodney, the Governor of Greenwich Hospital for transport down-river. Landed at Greenwich Hospital, it was carried about half a mile up the hill to the Royal Observatory. The box was opened and the watch wound and set going to mean time by Kendall. It was then placed in a deal box with a glass lid and two locks, which was screwed down to a seat in the east window of the Transit Room.[15] One of the keys was kept at the Royal Observatory, the other at Greenwich Hospital, so that the box could in theory not be tampered with unless both keys were present.

Every day about noon for the next ten months, in the presence of an officer from Greenwich Hospital who brought one of the keys up the hill with him, the box was unlocked. Maskelyne or the Assistant then wound the watch and compared it with the Transit Clock – which was checked against the heavens every day the weather permitted – recording the results in a special register, attested by the signatures of both parties.[16] The watch was then replaced and the box twice locked. Except for seven weeks when it was propped up at various angles for trial purposes, the watch remained horizontal, face up, lying on the same cushions and in the same inner box in which William Harrison had kept it on the voyage to and from Barbados.[17]

Having got these trials under way, Maskelyne needed to see the performance of the three large timekeepers which Harrison had failed to deliver as promised. On 23 May, therefore, accompanied by Dr Roger

Long from Cambridge, another Commissioner, and two watchmakers, William Frodsham and Walter Williams, Maskelyne arrived in Red Lion Square with an authorization from the Board to remove 'the three several Machines or Time Keepers now remaining in your hands which are become the property of the Public.'[18]

According to William Harrison's own manuscript journal, in a report purporting to have been signed by the two watchmakers, the former was remarkably uncooperative in advising how they should be transported; he did, however, say that the large one (presumably H1) should be removed whole, the other two in pieces; he then left the room:

> However Mr Maskelyne with disdain declared although Mr Harrison would not give any directions or Assistance that he knew as well as Mr Harrison how to convey the same, and about 4 o'Clock the said Mr Maskelyne in giving directions to the persons he had brought to carry the said Time Keepers or machines for putting the Machine in the large Frame a smith being sent for came accordingly and with a Pair of pincers in taking the same down, let the same fall and which we believe broke some of the movements therein at which Mr Maskelyne's hand was then on the Machine.
>
> Wm. Frodsham
Walter Williams[19]

Harrison later claimed that H1 was dropped by the workmen and that all three were further damaged by being taken to Greenwich on an unsprung cart instead of being carried by 'chairman's horse' (slung on two poles as with a sedan) as far as the river and then going by water.[20] Exactly what did happen, we do not know, but certainly Maskelyne had now replaced Bradley, Bliss and Morton as the Harrisons' Enemy No.1.

Once again, the Harrison affair had a quiescent period – for eight months. Then, on 27 January 1767, a Petition was presented to Parliament by 'several Merchants and other interested in the Navigation & Commerce of these Kingdoms.' They complained that, though the voyage proving the efficacy of H4 had finished as long ago as July 1764 and despite the award of considerable sums of public money, the principles and construction of the watch had still not been made public 'so as to be the least Benefit to Navigation'. Much worse, a French artist *had* obtained this knowledge and it was feared that the French would be the first in general possession of this great Discovery – which would be 'very detrimental to the Interest of these Kingdoms'.[21]

The report was referred to a select committee which included Captain Lord Howe (Treasurer of the Navy and a Commissioner of Longitude), the Chancellor of the Exchequer (Charles Townshend) and Captain Sir John Lindsay, of the *Tartar* in the Barbados trials. Exactly what were the politics behind this petition we do not know. Perhaps Harrison's friends, of whom we know only two for certain, James Short and Sir John Cust (the Speaker of the House of Commons and another of the Commissioners) had some hand in it. Be that as it may, the Petition stirred the Board of Longitude into activity. They met on 14 March and questioned Mudge on his disclosures to Berthoud a year earlier. In his evidence, he said he thought it was the Parliament's intention that he should disclose the information he had obtained at Harrison's disclosure. He had given similar information to a dozen or so English watchmakers as well as Berthoud.

A week later, Maskelyne presented his report on the Greenwich trials of H4. He summed up as follows:

> That Mr Harrison's Watch cannot be depended upon to keep the Longitude within a Degree in a West India Voyage of six weeks; nor to keep the Longitude within half a degree for more than a few days; and perhaps not so long, if the Cold be very intense; nevertheless, that it is a useful and valuable invention, and, in conjunction with observations of the Distance of the Moon from the Sun and fixed Stars, may be of considerable advantage to Navigation.[22]

The Board ordered his whole report to be printed. For a technical analysis of the results, the reader is referred to Quill (1966), Chapter 18, who concludes that, though on the face value of the results Maskelyne's summing up is fair and reasonable, he made no allowances in Harrison's favour: in particular, he could have made an allowance for the rate of going found at the beginning of the trial, which had changed considerably since the watch had been dismantled in 1765.

On 2 April 1767, the select committee gave a report to the House of Commons. To answer the complaints in the petition, they had examined William Harrison who said the drawings and explanation were to be published that very month.[23] On the French aspect, Mudge said much the same as he had already said to the Board, while Short said he had heard from Lalande and Lemonnier that Berthoud and Pierre Le Roy had been competing for a Premium recently offered in Paris for marine watches. Berthoud himself had told Short that, since the news of Harrison's award, every little watchmaker in France was endeavouring to make marine watches, 'and we shall see what this ferment will produce.'[24] The matter was referred to a Committee of the Whole House. However, before this could be resolved, the King prorogued Parliament and, when they reassembled in the autumn of 1767, the matter seems to have been forgotten – or ignored – and Parliament did not discuss the Harrison affair again until 1773.

The Board met again on 2 May, athe main business being the signing of an agreement with Kendall to make a copy of H4 for £400. Arrangements were made for sending H4 from Greenwich to London by water.

Maskelyne's *An Account of the going of Mr John Harrison's watch, at the Royal Observatory, from May 6th, 1766, to March 4th, 1767* ... was quickly published, swiftly followed by a vitriolic attack by Harrison in a booklet (price 6d) dated 23 June 1767 and entitled *Remarks on a Pamphlet lately published by the Rev. Mr. Maskelyne* ... (though Maskelyne's *Account*, published at 2s 6d, was hardly a pamphlet, consisting as it did of 84 quarto pages). In his *Remarks*, Harrison accused Maskelyne of all sorts of irregularities in the conduct of the trial – that the Greenwich Hospital officers were so decrepit that they could hardly climb the hill, let alone see whether Maskelyne was cheating; that the locks on the box could be picked with a crooked nail; that the box and the watch was exposed to the south east (and presumably the morning sun) whereas the thermometer was in a shady part of the room; even a hint that Maskelyne might have tampered with the transit clock. He accused Maskelyne of having drawn all the wrong conclusions from the data. He claimed that the lunar method was of very little use – less accurate, only available a few days in the month – despite the fact that it had already cost the public £6600 with no proper experiments. And, said he, Maskelyne had a special interest in that method 'in a pecuniary way'.

Maskelyne made no public reply. (Gould says he planned one, but gives no source.[25]) Though his, Maskelyne's, conclusions may have been a bit sweeping, both contemporary and present-day accounts refute most of the accusations.[26]

The minutes of the Board of Longitude's next meeting on 12 December 1767 contain no mention of Harrison's name. This had only happened twice before in all the thirty-seven Board meetings since 1735. Perhaps, therefore, now is a good moment to leave the Harrison affair for a while – with Harrison making his second longitude watch (H5) and Kendall making his copy (K1) of Harrison's first (H4) – and to continue in Chapter 11.

9

The Nautical Almanac

The annual *Nautical Almanac and Astronomical Emphemeris* and its companion *Tables Requisite to be used with the Astronomical and Nautical Emphemeris* – these were undoubtedly Nevil Maskelyne's greatest contribution to the improvement of navigation and astronomy – and to science as a whole. And it is largely because of the world-wide success of that publication in the eighteenth and nineteenth centuries that the world's system of longitude and time reckoning should today be based on the Greenwich meridian. It was almost entirely through Maskelyne's efforts and persistence that it came to be published in the first place – for the year 1767 – and it was he that became its first editor, superintending all the complex calculations whose precision was improved year by year as a result of the work of mathematicians and astronomers throughout Europe, with whom Maskelyne kept in close touch despite its bellicose state at that period. He was entirely responsible for the forty-nine issues of the almanac, from 1767 to that for the year 1815, published in 1811, the year of his death; and for three editions of the Requisite Tables, published under his aegis in 1766, 1781, and 1801.

As we have seen, Maskelyne's proposal that such an almanac should be published by the Board of Longitude was approved by Parliament in May 1765. By mid-June, the layout and content had been agreed – it was to be 'for the purposes of Astronomy as well as Navigation'[1] – and a sort of 'cottage industry' system of computing decided upon. All the important calculations were to be done quite separately by two 'computers' working independently in their own homes, sending their results to the Astronomer Royal, who would forward them to a 'comparer' for checking the one against the other. This system worked remarkably well during Maskelyne's time, very few errors creeping through. Maskelyne caught Reuben Robbins and Joseph Keech 'acting collusively' when computing the almanacs of 1771 and 1772: they were never employed again.[2]

He set out to give the navigator all he needed for reducing the observations taken at sea for finding position – for both latitude and longitude by any method, but particularly for longitude by lunar distance. 'All the lunar calculations for finding the longitude at sea by that method', he wrote to his brother in India a year after the decision to publish was taken, 'will be ready performed: & other useful & new tables added to facilitate the whole calculation; so that the sailers will have little more to do than to observe carefully the moon's distance from the sun or a proper star, which are also set down in the ephemeris, in order to find their longitude.' The ephemeris was also necessary for reducing the observation to find local time at the ship when using the chronometer method. 'The board of longitude', he continued, 'are also desirous to encourage the making of watches after Mr. Harrison's method. They have engaged a person to make one. I have had the drawings engraved here under my eye, & shall publish them in a short space of time.'[3]

On 13 June 1765, Israel Lyons the younger and George Witchell were engaged to compute for the almanac the first six months of 1767, William Wales and John Mapson the second six months, at a rate of £70 per annum per almanac (raised to £75 in December 1767). In July, Richard Dunthorne was appointed comparer and corrector at the same rate of pay. He and Maskelyne also began to prepare the *Tables Requisite to be used with the Astronomical and Nautical Ephemeris* described below.

At their meeting of 26 April 1766, the Board heard that Lyons and Witchell were falling behind in their computations for 1767 and it was agreed that Wale's and Mapson should stop work on 1768 for the time being, to help with computing 1767. Messrs Richardson & Clark of Salisbury Court, Fleet Street, were licensed to print the almanac and the Board's other publications; and John Nourse, bookseller in the Strand, and Messrs Mount & Page, stationers in Tower Hill, were licensed to publish and print the same, taking 20% of the selling price.

At last, on 6 January 1767 (that was the date stocks were received from the printer, although the imprint says 1766), just nineteen months after their compilation had been ordered by Act of Parliament, *The Nautical Almanac and Astronomical Ephemeris for the year 1767* and the *Tables Requisite* . . . were published by order of the Commissioners of Longitude. Though they had decided in 1765 that 3000 copies of the almanac should be printed, in fact only 1000 almanacs and 10 000 tables were actually printed, for sale at 3s. 6d. and 2s. 6d. respectively. By January 1784, 242 almanacs and 6992 tables remained unsold, so the Board's initial market research left something to be desired.[4]

Early in 1767, the Board published Mayer's last manuscript tables of the Moon more or less verbatim, *Theoria Lunae juxta systema Newtonianum*, *auctore Tobia Mayer*, seen through the press by Maskelyne. In March 1770, the Board published an edition of both his solar and lunar tables in Latin and English with an extensive introduction and explanation by Maskelyne.[5] This was followed in March by *The Principles of Mr. Harrison's Time-Keeper, with Plates of the same*, and in April by *An Account of the going of Mr. Harrison's Watch*...

In December 1767, the Board decided that the Nautical Almanac should be published three years in advance, and authorized Maskelyne to recruit additional computers, among whom were the infamous Robbins and Keech. This was an important decision because voyages of exploration could last several years. Captain James Cook, RN, in a passage in his journal full of praise for the almanac, was to say in 1770, '... but unless the Ephemeris is published for some time to come more than either one or two Years it can never be of general use in long Voyages and in short Voyages its not so much wanting...'.[6] On his first voyage (1768–71), Cook had the almanacs for 1768 and 1769 only; on his second (1772–5), he had 1772, 1773, 1774, and the few sheets that had been printed for 1775; on his third (1776–80), he ran out again, not having 1779 or 1780. Captain George Vancouver, on the other hand, for his voyage to the Pacific, the length of which no one could forecast before he sailed, was able to take a full set of almanacs to cover the necessary years 1791 to 1795.

The Nautical Almanac as planned by Maskelyne had twelve pages for each month, the headings of which are shown in Figs. 9.1 (*a*)–(*l*). As well as the usual monthly calendar giving saints' days and holidays, university terms, royal birthdays, etc., the places of the Sun were given daily at noon, of the Moon at noon and midnight, and of the five planets every seven days – all in *Greenwich Apparent Time* (as opposed to *Greenwich Mean Time*). There was also a table giving predicted times of the eclipses of Jupiter's satellites (an average of 17 for Satellite I, 8 for II, 8 for III, and 4 for IV, in any month where Jupiter was observable), and a series of diagrams showing the positions of these satellites relative to the planet every night at 3 a.m., Greenwich Apparent Time.

All this was quite standard information and could, for example, be obtained just as easily (for the meridian of Paris) from Lalande's *Connaissance des Temps*. Maskelyne's real innovations were, first, his tables of pre-computed lunar distances for the meridian of Greenwich, each month in the almanac as in Fig. 9.1 (*h*)–(*k*); and secondly, the Requisite Tables.

(a)

OCTOBER 1772. [109]

Days of the Month	Days of the Week	Sundays, Holidays, &c.	Phases of the Moon.
			D. H. M.
			First Quarter —— 3. 5. 8
			Full Moon —— 11. 5. 26
			Last Quarter —— 19. 5. 18
			New Moon —— 25. 21. 49
1	Th.	Remigius.	
2	F.		
3	Sa.		Other Phenomena.
			D.
4	Su.	16th Sunday after Trinity.	3. ☿ β ♍ diff. Lat. 11′.
5	M.		4. ☾ β ♑ 18ʰ. 6′.
6	Tu.	Faith.	7. ☾ θ ♒ 2ʰ. 55′.
7	W.		10. ☿ η ♍ diff. Lat. 25′.
8	Th.		

(b)

[110] OCTOBER 1772.

Days of the Month	Days of the Week	Sun's Longitude.	Sun's Right Asc. in Time.	Sun's Declin. South.	Equat. of Time. Sub.	Diff.
		S. D. M. S.	H. M. S.	D. M. S.	M. S.	S.
1	Th.	6. 8.48.58	12.32.23,3	3.29.57	10.34,5	18,5
2	F.	6. 9.48. 8	12.36. 1,3	3.53.15	10.53,0	18,2
3	Sa.	6.10.47.20	12.39.39,6	4.16.30	11.11,2	17,9
4	Su.	6.11.46,34	12.43.13,5	4.39.42	11.29,1	17,5
5	M.	6.12.45.50	12.46.57,3	5. 2.50	11.46,6	17,1
6	Tu.	6.13.45. 7	12.50.36,7	5.25.55	12. 3,7	16,8
7	W.	6.14.44.26	12.54.16,4	5.48.55	12.20,5	16,3
8	Th.	6.15.43.47	12.57.56,6	6.11.51	12.36,8	15,9
9	F.	6.16.43.10	13. 1.37,2	6.34.42	12.52,7	15,4
10	Sa.	6.17.42.35	13. 5.18,3	6.57.27	13. 8,1	15,0
11	Su.	6.18.42. 2	13. 8.59,9	7.20. 7	13.23,1	14,4
12	M.	6.19.41.30	13.12.41,9	7.42.41	13.37,5	14,0
13	Tu.	6.20.41. 0	13.16.24,4	8. 5. 8	13.51,5	13,5
14	W.	6.21.49.33	13.20. 7,5	8.27.28	14. 5,0	12,9
15	Th.	6.22.40. 8	13.23.51,1	8.49.42	14.17,9	12,2
16	F.	6.23.39.45	13.27.35,4	9.11.48	14.30,1	11,6
17	Sa.	6.24.39.25	13.31.20,3	9.33.46	14.41,7	11,0
18	Su.	6.25.39. 7	13.35. 5,8	9.55.36	14.52,7	10,1
19	M.	6.26.38.51	13.38.51,9	10.17.17	15. 3,1	
		28.38	13.42			

(c)

OCTOBER 1772. [111]

Days	Semidiameter of the Sun.	Time of D° passing the Meridian.	Hourly Motion of the Sun.	Logarithm of the Sun's Distance.	Place of the Moon's Node.
	M. S.	M. S.	M. S.		S. D. M.
1	16. 3,0	1. 4,3	2.27,9	9.999912	6.20.16
7	16. 4,6	1. 4,6	2.28,5	9.999141	6.19.57
13	16. 6,3	1. 5,1	2.28,9	9.998388	6.19.38
19	16. 7,9	1. 5,6	2.29,3	9.997671	6.19.18
25	16. 9,6	1. 6,2	2.29,8	9.996981	6.18.59

Eclipses of the SATELLITES of JUPITER.

	I. Satellite. Emersions.		II. Satellite. Emersions.		III. Satellite.
Days	H. M. S.	D.	H. M. S.	D.	H. M. S.
2	18.21. 7	1	13.11.20	4	6°25.23 I
	12.50.35	5			9°55.33 E

(d)

[112] OCTOBER 1772.

Days	Heliocentric Longitude.	Heliocentric Latitude.	Geocentric Longitude.	Geocentric Latitude.	Declination.	Passage over Merid.
	S. D. M. S.	D. M.	D. M.	D. M.	D. M.	H. M.
			MERCURY.	Greatest Elong. 6°.		
1	1.23.40	0.58 N	5.23. 8	0.22 N	3. 4 N	22. 2
7	3. 1.22	5. 0	5.26.47	1.35	2.44	22.58
13	4. 7. 2	6.55	6. 4.34	0. 0	0. 0	23. 6
19	5. 7.27	6.30	6.14. 9	1.52	3.52 S	23.18
25	6. 2.33	4.47	6.24.11	1.24	8. 5	23.32

(e)

OCTOBER 1772. [113]

Days of the Month	Days of the Week	Moon's Longitude at Noon.	Moon's Longitude at Midnight.	Moon's Latitude at Noon.	Moon's Latitude at Midn.
		S. D. M. S.	S. D. M. S.	D. M. S.	D. M. S.
1	Th.	8.10.51.10	8.17.50.25	4. 6.50 N	4.29.26 N
2	F.	8.24.43.19	9. 1.30. 1	4.47.42	5. 1.35
3	Sa.	9. 8.10.42	9.14.45.35	5.11. 9	5.16.30
4	Su.	9.21.15. 2	9.27.39.23	6.17.41	5.14.53
	M.	10. 3.50			4.58. 3

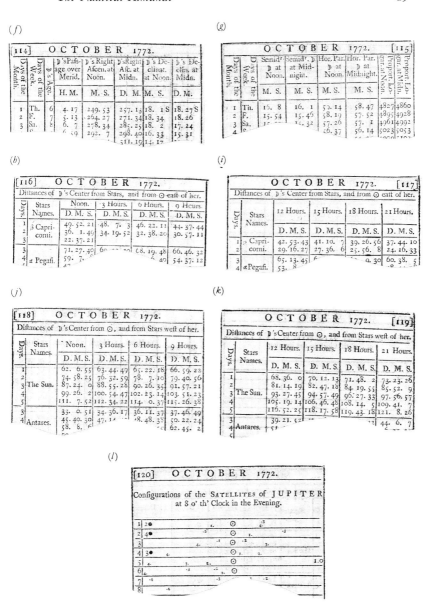

Fig. 9.1. The Nautical Almanac as designed by Maskelyne. For each month there were twelve pages of data, the headings of which are shown here. 'S.D.M.S.' signifies Sign of the zodiac, Degrees, Minutes, and Seconds of arc; 'D.M.S.' signifies Degrees, Minutes, and Seconds of arc; 'H.M.S.' signifies Hours, Minutes and Seconds of Greenwich apparent time (not Greenwich mean time). (Photos: RGO)

In the almanac, lunar distances to the Sun were tabulated every three hours whenever that distance was between 35° and 120°, roughly when the *Moon's age* was between 4 and 10 days, and again when it was between 19 and 26 days, a total of about 15 days in each lunar month; on those same days, the distance to at least one star was tabulated also. On the 13 days when the Sun and Moon were too close or too far apart to be used, lunar distances to at least two stars, one on each side of the Moon, were tabulated. Ten stars only were used: first-magnitude Aldebaran, Pollux, Regulus, Spica, Antares, Alpha Aquilae (Altair), and Fomalhaut; second-magnitude Alpha Arietis (Hamal) and Alpha Pegasi (Markab); and third-magnitude Beta Capricorni.

These tables did away with one of the most time-consuming and difficult parts of the reduction of a lunar-distance observation: calculating the Moon's place from Mayer's tables and then computing the distance between the two bodies in arc-seconds by logarithms (in fact, this degree of accuracy was not really justified by the data[7]), which, said Maskelyne, 'are the principal and only very delicate Part of the Calculus'.[8] But with the almanac, the navigator could obtain the Greenwich Apparent Time of his observation merely by comparing his true lunar distance with the quantities tabulated every three hours in the lunar-distance table, and making a simple interpolation.

The idea for these tables was not originally Maskelyne's but came from Lacaille, who had successfully used the lunar method in the 1750s on his voyage to and from the Cape. In the *Connaissance des Temps pour l'Année 1761* (published in 1759), Lalande published a paper by Lacaille entitled 'Méthode pour trouver les Longitudes en mer par le moyen de la Lune'. Lacaille included sample pre-computed lunar distance tables for every four hours on the meridian of Paris for fourteen days in July 1761: he promised more tables in future years, but these did not materialize. F. Marguet, the French historian of navigation, continues the story:

> But to execute this project, he [Lacaille] had neither the time nor the money. England forestalled us in doing this, and, there again, the transit of Venus of 1761 was the occasion of progress of longitude at sea. Maskelyne went to St Helena to observe the transit. He did on the voyage what Lacaille had done ten years before and he took the advice of the latter. As early as 1763, he proposed, in his *British Mariner's Guide* – a work which appears to have been very little distributed with us, since Rochon could not procure one for his voyages – to adopt the French astronomer's plan. He did not leave it there but obtained, after much tenacious lobbying, the publication of the almanac whose form had been given in France. . . The work, whose aim was to contribute greatly to the improvement

of astronomy, geography, and navigation, was to be published annually – and this actually occurred...[9]

Maskelyne's other innovation was to provide the Requisite Tables. With the first edition of these, all the navigator needed to reduce his observations were the Requisite Tables and a set of simple trigonometrical tables. With the second and third editions, he needed to have only one book, because all the necessary trigonometrical tables were included. The first edition comprised:

- (a) all those tables that did not change from year to year (refraction in altitude, parallax in altitude, time and arc, dip of sea horizon);
- (b) star places, which change very slowly;
- (c) tables of Proportional Logarithms, not included in ordinary log tables; and
- (d) two alternative versions of tables, devised by Israel Lyons and Richard Dunthorne respectively, for 'clearing the distance', to correct the measured lunar distance for two effects:

    - (i) *atmospheric refraction*; because of this, heavenly bodies appear to be higher above the horizon than they really are, the lower the body, the greater being the effect: a body apparently at $10°$ altitude is actually at $9°55'.3$, one at $30°$ is actually at $29°58'.3$;
    - (ii) *parallax*; the monthly lunar-distance tables were calculated assuming the observer to be at the centre of the Earth, whereas he is actually at its surface (in fact, a few feet above the surface – 19 feet height of eye on the *Princess Louisa*'s quarter deck, for example[10] – which effect is allowed for by the dip table in (*a*) above: the effects of parallax depend, first, upon the body's altitude; secondly upon the body's distance from Earth – for stars at an infinite distance there is no parallax effect, for the Sun the effect can generally be ignored except at low altitudes, for the Moon it is very significant in this context.

We will conclude this chapter by giving a brief description of the navigator's procedure for finding his position at sea using the new almanac:

For latitude
Observe the *Meridian altitude* of the Sun – the 'noon sight' – correct for

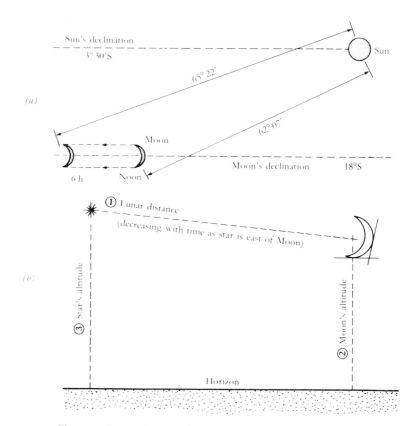

Fig. 9.2. Lunar Distance Observations.
(a) How lunar distance changes with time. The figures are for noon and 6 p.m., Greenwich apparent time, on 1 October 1772, from the almanac pages 110, 114, and 118 in Fig. 9.1.
(b) Showing the three near-simultaneous observations needed to find longitude by lunar distance:
(1) the angular distance between the Moon and a selected star (or Sun);
(2) the altitude of the Moon above the horizon; and
(3) the altitude of the selected star (or Sun).

dip, refraction, etc., and apply the Sun's declination obtained from the almanac. Latitude can also be obtained by observing the Pole Star, though Maskelyne did not include the necessary tables in his almanacs.

For longitude by lunar observations

Make three more or less simultaneous observations (preferably with at least two observers), one lunar distance and one altitude of each of the two bodies (see Fig. 9.2), timed with a pocket watch (not necessarily a chronometer). With the Sun, observations were taken in daylight when both bodies were at least 10° above the horizon. With a star, morning or evening twilight, when the horizon is easily visible, was preferable; if not, a separate altitude observation to find local time had to be made when the horizon *was* visible, and a *running fix* used. The procedure for reduction was briefly:

(a) Find the Local Apparent Time of Observation from the measured altitude of the Sun or star, which must be at least two hours from the meridian. Obtaining the Sun's place from the almanac, or the star's place from the Requisite Tables, solve a spherical triangle by trigonometry.

(b) Clear the measured lunar distance of the effects of refraction and parallax, using Lyons's or Dunthorne's tables, to obtain true lunar distance.

(c) From the true lunar distance, find the Greenwich Apparent Time of observation by interpolation in the tables in the almanac. (Figs. 9.1 (*h*)–(*k*)

(d) The difference between Local Apparent Time in (a) and Greenwich Apparent Time in (c) gives the longitude.

For longitude by chronometer

Observe the altitude of the Sun or a star at least two hours from the meridian, timing by chronometer (or by a *deck watch* which is immediately compared with the chronometer). Carry out Step (a) above, then convert Local Apparent Time to Local Mean Time by applying the *Equation of Time* given in the almanac. Longitude is found from the difference between the Local Mean Time so obtained, and the Greenwich Mean Time from the chronometer whose error is known – or at least estimated.

In 1768, Richard Bishop, a sea pilot examined by Parliament in 1763 and now a London navigation teacher recognised by the Board of Longitude, published a *pro forma*[11] which made the lunar calculations easier. A real-life example is shown in Fig. 9.3.

The first Nautical Almanac was published on 6 January 1767 – without doubt this was the most important date in the history of the art of navigation, certainly since the invention of the reflecting quadrant thirty-six years earlier, perhaps since the beginnings of latitude navigation back in the fifteenth century.

Before that, probably not more than a score of navigators of any nationality had succeeded in measuring their longitude when out of sight of land. Now, with the publication of Britain's Nautical Almanac, any competent mariner could do so quickly and comparatively easily, provided he could afford a Hadley quadrant at about £5 (a new-fangled brass sextant would be much better though far more expensive at £15), a watch accurate to one minute in six hours, a book of tables (say Robertson's *Elements of Navigation*, two volumes for 15s.), and the almanac and requisite tables at 6s. the pair. Of course, when he had one or more reliable chronometers (at least forty guineas each), he would find it easier still – but that was to be at least forty years ahead for all but the favoured few.

Though the means were there, it naturally took many years for the practice to become common-place, because mariners are a conservative breed. In Britain, however, the officers of the East India Company and of the Royal Navy showed the way. Some of the former had already begun to find their longitude the hard way, using the methods described in *The British Mariner's Guide*: now they thankfully turned to the new almanac. In the Royal Navy, the almanac and requisite tables were supplied to many of HM Ships in 1767, and, early in 1769, an order went out to the Fleet that the masters of all ships visiting Portsmouth – the master was responsible for a naval vessel's navigation, under the orders of the captain – were to go to the Royal Naval Academy for instruction in the use of the Hadley quadrant and the almanac, obtaining a certificate accordingly, the Board of Longitude paying the staff there half a guinea per pupil. And the Navy Board was told that no one should in future be appointed master of one of HM Ships unless he could produce such a certificate from the Academy, or from Robert Bishop or Samuel Dunn, navigation teachers in London approved by the Board.[12] Alas, a combination of conservatism and war conditions made it impossible for the Royal Navy to enforce these idealistic orders, as we shall see in Chapter 11.

Fig. 9.3. The Computation of an actual Lunar-Distance Observation for Longitude, made on 4 October 1772 (using the almanac pages in Fig. 9.1 above) in an unknown ship, probably an East Indiaman, some 500 nautical miles (930 km) west of the Canary Isles. This is a page from a book of forms recording observations taken from 7 to 25 June 1772 when the ship was approaching the Cape of Good Hope from the Indian Ocean, and from 4 to 19 October in the Atlantic. It ends with an observation in the mouth of the Straits of Dover on 19 November. The navigation teacher, Robert Bishop, published these forms in 1768, for use with the newly-published Nautical Almanac. From Board of Longitude papers, RGO MS.14/67. f.46v. (Photo: RGO)

The Nautical Almanac

17. 4"

Computations for finding the Longitude by Observations taken 9ᵗʰ Octʳ 1772.

Time by Watch	Dist ☉ or ☾	Altitude ☉ or ☾	Altitude ☾
1. 20. 15	102. 26. 45	19. 14	31. 40
1. 26. 0	102. 27	18. 00	32. 42
1. 32. 15	102. 24	16. 40	32. 44
1. 39	102. 20. 45	12. 02	33. 1A
1. 22. 20	102. 26. 33	18. 01	32. 43
Mean	Mean	Mean	Mean

For rectifying the Watch Observ'd at 1. 26. 20

Altitude ☉ lower limb po. 01′ Watch by

Suppos'd Apparent time

Latitude compᵈ 27. 20 .... Longitude compᵈ 91

—From 90. Co Lat. 62. 33 ..... + or — in time =p. 6 ̕ ̕ . 40

Time Observ'd by Watch 1. 26. 20

Suppos'd Apparent time at Greenwich 6. 14. 20

For Computing the Apparent time.

	h ̕ ″	̊ ̕ ″
☉ Declination	3. 4. 39. 42	4. 09. 42
Dᵒ	3. 5. 2. 30	
Difference in 24 hours 23. 88	=	6
☉ Declination at the time of Observation 4. 15. 42		
+ or — from 90. is Polar Distance 94. 46		
Altitude ☉ lower limb Observ'd 18.01		
+ ☉ Semidiameter less by dip & Refraction 7		
Altitude ☉ center corrᵈ 18. 30		
—From 90. is Zenith Distance 71. 52		

Zenith Distance 71. 52

Polar Distance	94. 46	Ar.Compᵗ Sine	00150
Co Latitude	62. 30	Ar.Compᵗ Sine	05150
Sum	229. 16		
½ Sum	114. 30	= 65. 22 Sine	95056
½ Sum — Zenith Distance	= 42. 36 Sine		98100
Sum			1041340
½ Sum is Co Sine	38. 22		89174
Doubled	2		
Horary ∠	66. 44	p. 6.	4. 26. 16
Time by Watch when the Altitude ☉ was taken	1. 26. 20		
Difference is Watch		by	126

To Compute from the Observations above

Time by the Watch when Dist ☾ was taken 1. 26. 20

Watch being .... by 26

Apparent time at taking the Distance ☾ 1. 26. 16

+ or — for Longitude from Greenwich compᵈ

Apparent time at Greenwich

Mean of the Observ'd Altitude ☉ or 18. 01

+ Semidiameter less by Dip 10

Altitude of the ☉ or . corrᵈ 18. 11

Mean of Observ'd Altitude ☾ 32. 43

+ or — Semidiamᵗ according which limb is Observ'd .. 01

Altitude of ☾ corrᵈ 32. 24

By the Ephemeris.

	̕ ″	̕ ″
Semidiameter ☾ at	15. 26	15. 26
Dᵒ at	15. 20	
12 hours Difference	6	= 3
		15. 23
+For Increase of Altitude ☾	p. 153.	0
Apparent Semidiameter ☾		15. 8
+ Semidiameter		16. 4
Sum of Apparent Semidiameter ☉ & ☾		31. 36

Computations continued

Horᵃ Parˢ at 16. 37′ Propˡ Logarithm 5045

At 16. 14 Dᵒ 5053

Propˡ Logarithm ...... 5045 12ᵈ Difference 60

Propˡ part + or — .... 15 Dᵒ is Propˡ part 15

Propˡ Logarithm 5048 = Horˡ Parallax

Distance ☉ or . & ☾ Observ'd 102. 26. 5

+ or — Semidiameters ☉ & ☾ 31. 36

Distance of the ☉ or . and ☾ centers 102. 58. 30

Computation of Refraction by Mʳ Lyons Table.

Alt ☽ corrᵈ	10. 1	32. 10	T.N.	1136	Dᵒ. T.N.	1436
Alt. Dᵒ.	32.45	33.10		1196		1031
T.N.			1156	1ˢᵗ Diffᶜᵉ	60	2ᵈ Diffᶜᵉ	105
+1ˢᵗ Propˡ part	41	1ˢᵗ Proˡ part	41	2ˢᵗ Proˡ part	2

Sum 1181

—2ˢᵗ Propˡ part 2

to this .... 1179 Number +for an Index 2. 1179

+Logarithm Co Secᵗ Distance 0112

Logarithm .... 435 .... 2. 1291

By Tab. 2ᵈ with Diff. & less. Alt. 26 Distance less 90 — more .

Sum or Difference is the ... 161 = 2. 41 Effect of Refracᶜ

Distance of ☉ or . & ☾ centers 102. 58. 30

Effect of Refraction 2. 41

Distance clear'd of Refraction 103. 01. 11

For Parallax

	̊ ̕ ″		
Altitude ☉ or . corrᵈ	18. 11		
—Refraction p. 2	3		
Alt. ☉ or . corrᵈ	18. 8	Co. Secᵗ	10. 5069
Dist ☉ or . & ☾ clear'd Refᶜ. 103. 1		Sine	9.9007
Propˡ Log Horˡ Parallax	3030		
Propˡ Log. Arch 1ˢᵗ	18. 9		.9994
Altitude ☾ corrᵈ	32. 24		
—Refraction p. 2	1		
Alt. ☾ corrᵈ	32. 23	Co. Secᵗ	10. 2712
Dist ☉ or . & ☾	103. 1	Tangᵗ	10. 6360
Propˡ Log Horˡ Parallax			3030
Propˡ Log. Arch 2ᵈ	6. 39		1. 4110
Arch 1ˢᵗ	18. 9		
Prinˡ Effect of Parallax	25. 01	or Parallax in Distance	
Distance clear'd of Refraction			103. 1
Petiᵗ Effect of Parallax			25. 1
Distance clear'd of Principal Effect of Parallax	102. 36. 10
By Table 4ᵗʰ for second corrᵗ of Parallax			2
Reduced Dist clear'd of Refraction & Parallax	102. 36. 8

By the Ephemeris

	̊ ̕ ″	̊ ̕ ″
Dist ☉ or . & ☾	A. at 6 102. 23. 14 Dᵒ	102. 23. 14
Dᵒ. at	103. 54. 23 Redᵈ Dist	102. 36. 0
in 3 hours	1ˢᵗ Diffᶜᵉ 1. 36. 9 2ᵈ Diffᶜᵉ	12. 54
Proportional Logarithmᵐ of 1ˢᵗ Difference	2634
Dᵒ	2ᵈ Difference	1447
Proportional Log.	26. 69 Diffᶜᵉ	0018
+hour of the 1ˢᵗ Dist.	6	
Gives Apparent time	6. 23. 2 at Greenwich
Apparent time	1. 26. 16 at taking the Dist ☉ or . & ☾
Difference	4. 56. 45 in time = p. 6. 74. 11
is Longitude between the Place of Observation and Greenwich

N.B. Distance clear'd of Refraction ⎰ less 90. take the Diffᶜᵉ of the two Arches ⎱ is Principal Effect of Parallax ⎰ Arch first greatest — centre + ⎱
⎱ More the Sum of the two Arches ⎰ ⎱ —from the Dist clear'd of Refᶜ ⎰

By the Requisite Tables find the Parallax in Alt ☾ by Table 4ᵗʰ with Distance & Parallax in ⎰ Alt ☾ ⎱ Differenceᶜᵉ secᵈ corrᵗ Parallax
⎱ Dist. ⎰

which is to be + if distance is less 90. but more —. By Robert Bishop

By the requisite Tables page 345, with App. Altitude ☾ and Horˡ Parallax, gives the Parallax in Altitude.

Publish'd according to Act of Parlᵗ Decemʳ. 9ᵗʰ 1768.

To close this first part of the book, we will quote from the conclusion of another book, Eva Taylor's magisterial history of navigation, *The Haven-finding Art*:

And the longitude being solved, there was no longer justification for any checks on the improvement of the sextant, or for any carelessness in the determination of latitude and local time. Maskelyne's first Nautical Almanac appeared in 1767, and with Hadley and Harrison what may be termed the pre-scientific age of navigation was brought to a close. Landmark or no landmark, the sailor knew precisely where he was – or had the means to know. He did indeed at long last possess the Haven-finding Art.[13]

Part 2

ACHIEVEMENT

10

Early years at Greenwich, 1765–9

Maskelyne's first years at Greenwich were very busy. Already a member of the Board of Longitude, he was elected to the Council of the Royal Society on 1 December 1766, remaining on Council, except for two short breaks, until his death forty-five years later. Membership of these two bodies gave rise to an enormous amount of work for the Astronomer Royal – the planning of the scientific side of the various voyages of exploration, the assessment of longitude proposals, the trials of chronometers, the refereeing of scientific papers, and many other matters. The editing of the Nautical Almanac was very time-consuming. But all this had to be fitted in with the observatory's fundamental task – observing, continuing his predecessors' work of providing the raw material for 'rectifying the Tables of the Motions of the Heavens, and the Places of the Fixed Stars'.[1]

So far, the chapters in this book have generally had a fairly firm subject basis. From here, however, so as to preserve chronological order yet at the same time tell of Maskelyne's many-sided activities in a coherent manner, the story will be told decade by decade, each chapter being divided into sections covering Maskelyne's main activities – personal and family affairs, the observatory and astronomy generally, the Board of Longitude and Nautical Almanac, and the Royal Society – generally presented in that order.

Personal and family

When Nevil Maskelyne became Astronomer Royal on February 8 1765 at the age of 33, his close relatives were these:

- uncle, Nevil Maskelyne, 73, unmarried, of the Down, Purton;
- uncle, James Houblon Maskelyne, 64, secretly married with two infant daughters, of Marlborough and Newbury;
- aunts, Jane and Sarah Maskelyne, 72 and 70, unmarried;
- brother, Rev. William Maskelyne, 40, unmarried but with a

putative daughter, a man of business of The Ponds, Purton Stoke;
- brother, Capt. Edmund Maskelyne of the East India Company service, 37, widower, his wife Miss Floyer of Abergavenny (possibly daughter of a former Governor of Madras) having died within a few months of marriage in 1762: he had purchased Basset Down House, on the Marlborough Downs a few miles south of Purton from a distant cousin in 1763;
- sister, Margaret, 30, married to Robert, Baron Clive of Plassey (see Chapter 2), of Berkeley Square, London.

In June 1764, Robert Clive had sailed for India to become Governor of Bengal, leaving his wife and young family in England but taking his brother-in-law Edmund as ADC. Nevil himself lived in Flamsteed House with a housekeeper and manservant. His assistant Joseph Dymond had rooms in the New Observatory (today's Meridian Building across the courtyard) but presumably dined at his table.

In a letter to Edmund in India dated 15 May 1766, Nevil tells him that brother Billy visits him occasionally 'but he is a man of business and travels a good deal'; that attempts to reconcile the dispute between brother Billy and Uncle Nevil (unspecified but presumably something to do with Maskelyne property) can never succeed; that the two Maskelyne aunts are dead; and thanks him for the gift of £200 of India stock.[2] Margaret, writing to Edmund the following day, says she often sees Nevil who is happy at Flamsteed House, and that they are to be neighbours that summer because she is renting Westcomb Park, a large house about a mile west of the observatory, for a year from that summer.[3] Nevil's letter has a tantalizing finish:

I shall be obliged to you for a little callico on your return, if you have room; a little that would suit a lady may possibly be useful some time or other.[4]

One wonders for whom the calico was intended: Nevil was not to be married for another eighteen years. But a quilted astronomer's 'observing suit' still survives – coat, waistcoat, and trousers (Fig. 10.1) – which was presumably brought back when Clive and Edmund did return from India in July 1767.

Since returning from Barbados, Nevil had dined fairly frequently as a visitor at the Mitre Club before Royal Society meetings as guest of various members, as did Dr Benjamin Franklin who was very active in the Society at that time. On 26 September 1765, John D. Ross, DD, (afterwards Bishop of Exeter) proposed Maskelyne as a member but no vacancy occurred immediately. At the Anniversary Dinner of 30 July 1767, how-

Early years at Greenwich, 1765–9

Fig. 10.1. Nevil Maskelyne's Observing Suit, quilted silk waistcoat, jacket and trousers, sent by Robert Clive from India, probably about 1765, now in the possession of Nigel Arnold-Forster, Esq. (Photo: NMM)

ever, the Club decided that the Astronomer Royal and the two Secretaries of the Society should henceforth be considered members *ex officio*. He dined there twelve times in 1765, five in 1766, seven in 1767, eighteen in 1768, and nineteen times in 1769. The menu on the day he was elected was as follows:

Fresh salmon	Skate	
Knuckle of veal, &c	Achbone of Beef	Veal cutlets
Goose roast	Sweetbreads	Pease
Damson tart	Butter & cheese	

All washed down with porter or lemonade, with claret at two shillings extra.[5]

In 1768, he took the degree of Bachelor of Divinity at Cambridge. For this the candidate had to 'perform an Act' and preach a sermon in Latin. The proceedings in divinity were held in high esteem, the Opponents were formidable, and the Regius Professor of Divinity acted as Moderator. Unlike the MA degree, the divinity exercises for both BD and DD were held in public, and attracted large audiences, especially for the more popular candidates. A bachelorship in divinity was certainly not easily won.[6]

The observatory

Maskelyne and his Assistant, Joseph Dymond, started making the regular observations described in Chapter 7 above on 7 May 1765. William Bayly took Dymond's place on 16 November 1766. Maskelyne had twenty-six assistants during the forty-six years he was Astronomer Royal, only three of whom stayed more than five years, and some of whom stayed only a few months. This high turnover should not be attributed to difficult master-apprentice relations – far from it – but rather to the very low rate of pay. Most came as young men, stayed a year or so learning astronomy from the master, both observing and computing, then moved on to more remunerative employment.

When in October 1765 the Board of Ordnance eventually agreed to pay for the repair and renewal of instruments, Maskelyne arranged for the work to be put in hand, James Short repairing his own 6ft Newtonian telescope, John Bird and Peter Dollond dealing with the other instruments, and John Shelton with the clocks. In July 1767, the King gave orders that the Board should be responsible also for paying up to £60 a year for the printing of the observations which Maskelyne, unlike previous Astronomers Royal, delivered annually to the Royal Society.[7] Though actually printed year by year, they were not published until 1776 when ten years' worth had accumulated.

On 2 June 1768, the King and Queen paid their first visit to the observatory.[8]

It was at this time that Maskelyne began the exchange of letters with

Obſerved TRANSITS of the FIXED STARS and PLANETS over the MERIDIAN. 55

In the YEAR MDCCLXXVI.

DAY of the MONTH.	1ſt Wire. M S	2d Wire. M S	3d or Meridian Wire. H M S	4th Wire. M S	5th Wire. M S	Names or Characters of the Stars and Planets.
☽ SEPT. 30	59,1	31,3	14 0 4,1	36,6	9,3	*Arcturus.*
	40,8	15,1	15 19 49,6	24	58,2	*α Cor. borealis.*
	51	21,5	15 27 52,1	23	53,7	*α Serpentis.*
	12,6	46,3	16 10 20,5	54,2:	28,3	*Antares.*
	8,1	39,2	17 19 10,5	41,6	13	*α Ophiuchi.*
	13	44	19 30 15	46	16,9	γ } *Aquilæ.*
	28	58,8	19 34 29,5	0,4	31	α } *Aquilæ.*
	55,8	26,5	19 38 57	27,5	58,3:	β } *Aquilæ.*
	49,1	20,5	19 59 51,8	23	54,3	1 α } *Capricorni.*
	13	44	20 0 15,5	46,7	18	2 α } *Capricorni.*
	54,5	25	21 48 55,5	26	56,5	*α Aquarii.*
	13,4	44,9	22 48 16,1	47,5	19	*α Pegaſi.*
♂ OCT. 1	The Pendulum vibrates 1° 26′ on each Side.					
		42,4	20 28 25,5	8,2	. . . . .	*α Cygni.*
	20,1	54,5	23 51 29	3,5	37,8	*α Andromedæ.*
	18,7	50	23 56 21,4	52,8	24,2	*γ Pegaſi.*
	36 10,8:	36 43,3	4 37 16,2	37 48,7	38 21,9	☽ 2 L. 15ʰ 56′ 58″,2 mean Time.
☿ 2	28 35	29 5,4	12 29 35,7	30 6,6	30 37,3	☉ 1 L.
	30 44,3	31 14,3	12 31 45,3	32 15,5	32 46,3	☉ 2 L.
	38,6	23 17,5	18	. . . . .	25 14,5	*α Lyræ.*
	25,5	56,4	19 34 27,1	58	28,9	α } *Aquilæ.*
	53,5	24	19 38 54,7:	. . . . .	. . . . .	β } *Aquilæ.*
		41:	20 28 24	7:	*24,7	*α Cygni.* *49″,7
♃ 3	52,1	22,7	19 38 53,4	24	54,5	*β Aquilæ.*
	57,6	40,2	20 28 23:	6:	48,5	*α Cygni.*
	50,6	21,3	21 48 51,9	22,2	52,8	*α Aquarii.*
	9,9	41,4	22 48 12,8	44	15,5	*α Pegaſi.*
	18	52,5	23 51 27	1,3	35,7	*α Andromedæ.*
	16,4	47,6	23 56 19	50,5	22	*γ Pegaſi.*
♀ 4	36,2	15	18 23 54	33	12	*α Lyræ.*
	8,2	39	19 30 10,2	41	12	γ } *Aquilæ.*
	23,1	54	19 34 24,8	55,6	26,5	α } *Aquilæ.*
	51	21,5	19 38 52,1	22,9	53,4	β } *Aquilæ.*
	8	39,1	20 0 10,6	42	13,2	*2 α Capricorni.*
	49,7	20,1	21 48 50,6	21	51,5	*α Aquarii.*
	8,6	40	22 48 11,2	42,7	14	*γ Pegaſi.*
			7 22 6,7:	37,1	8	*Procyon.*
	58,3	33	7 26 7,6	42,3	17	*Pollux.*
	28 57,2	. . . . .	7	30 37,1	. . . . .	{ ☽ 2 L. 18ʰ 37′ 33″,7 mean Time, cloudy, not obſerved at Quadrant.
☉ 6			12 44 6,7	44 37,5	45 8,3	☉ 1 L.
	45 14,4	45 45,3	12	46 46,6	47 17,4	☉ 2 L.
	52,2	24,7	14 59 57,2	29,7	2	*Arcturus.*
	33,9	8	15 19 42,5	16,8	51	*α Cor. borealis.*
	1	32,1	17 19 3,4	34,7	6	*α Ophiuchi.*
	21,2	51,8	19 34 22,9	53,1	24,7	*α Aquilæ.*
	5,9	37,2	20 0 8,5	39,5	11,2	*2 α Capricorni.*
	47,5	18	21 48 48,3	18,9	49,6	*α Aquarii.*
	6,7	37,7	22 48 9,4	40,7	12,1	*α Pegaſi.*
	14,5	49	23 51 23,5	57,9	32,4	*α Andromedæ.*
	13	44,2	23 56 15,8	47,2	18,7	*γ Pegaſi.*
	3,3	33,9	7 22 4,5	35	5,7	*Procyon.*
	56,4	30,8	7 26 5,5	40,2	15	*Pollux.*
	14 37,5:	15 10:	9 15 42,4:	16 15,4:	16 47,5:	☽ 2 L. 20ʰ 15′ 5″,2 m. T. very hazy.
	52,6	23,8	9 50 55,1	26,1	57,5	*Regulus,* very hazy.
☽ 7	46 44,1	47 14,8	12 47 45,3	48 16,2	48 46,8	☉ 1 L. } hazy.
	48 53,5	49 24	12 49 54,7	50 25,1	50 56	☉ 2 L. } hazy.
	5,3	47 36,6	22	48 39,5	. . . . .	*α Pegaſi.*
	13,2	47,7	23 51 22,3	. . . . .	31,4	*α Andromedæ.*
	12	43,4	23 56 14,6	46	17,5	*γ Pegaſi.*
			7 22 3,4	34	4,7	*Procyon.*
	55	29,6	7 26 4,5	39,1	13,8	*Pollux.*

Fig. 10.2. Published Transit Results for 1776. From NM (1787b), *Greenwich Observations 2*, Transits, 55. 'H.M.S.' signifies Hours, Minutes and Seconds of sidereal time. (Photo: RGO)

foreign astronomers and mathematicians which was to prove so fruitful. In April 1767, for example, he wrote in Latin to Father Boscovich, by now Director of the observatory in Pavia, telling him of the success of Peter Dollond's 3½ft triple achromatic telescope mentioned in Chapter 7 above. Boscovich immediately ordered one from Dollond, and the arrangements for its delivery were made by their mutual friend Alexander Aubert, an amateur astronomer of distinction who had many mercantile contacts in northern Italy.[9]

In March 1769, he was visited by Jean Bernoulli III, Astronomer Royal to Frederick the Great of Prussia; Bernoulli published an interesting description of what he saw at Greenwich.[10]

Another interesting 'foreign' contact was Dr John Ewing of Philadelphia who, in 1769, wrote to Maskelyne and to Franklin, pointing out the potential advantages of having an observatory in Philadelphia, extolling 'the serenity of our air'. He asked Franklin to recommend this to the American Philosophical Society, suggesting that Maskelyne might 'afford some assistance from home', proposing that the Philadelphia observatory should be 'made subservient to his observatory at Greenwich and put under his general direction.'[11] No action was taken on this interesting suggestion, though we shall hear of it again.

The most important astronomical event in the period covered by this chapter was the second transit of Venus, predicted to occur on 3 June 1769, world-wide preparations for which (in which Maskelyne himself was closely concerned) will be considered later. One effect of these preparations was that Rev. Malachy Hitchins, already working for Maskelyne as Nautical Almanac comparer, came to the observatory in April 1769 as a temporary replacement for William Bayly, who was to observe the transit at the North Cape of Norway.

At Greenwich, the start of the transit did not occur until after 7 p.m. on 3 June, so the Sun was very low and conditions were far from ideal despite a fine day. Nevertheless, there was a distinguished gathering of observers:

> *In the Great Room*
> The Astronomer Royal, observing with the 2ft focus Gregorian reflecting telescope by Short that Charles Green had used at Greenwich for the transit of 1761;
> John Horsley, who had first met Maskelyne when a ship's officer in the *Warwick* on the voyage from St Helena in 1762, and who

had since made several more voyages to the East Indies and taken many lunar observations for longitude, using the observatory's 10ft achromatic refractor;

Samuel Dunn, a teacher of mathematics and navigation in London, using a 3½ft achromatic refractor.

In the Eastern Summer House
Malachy Hitchins, with the observatory's 6ft Newtonian reflector by Short;

Rev. William Hirst, a former naval chaplain who, as chaplain to the garrison in Madras, had observed the 1761 transit there, observing at Greenwich with a 2ft reflector by Nairne (belonging to Dunn), assisted by Henry Vansittart, former Governor of Bengal: both men were lost when the *Aurora* frigate disappeared early in 1770 on passage to India.

In the Western Summer House
Peter Dollond, optician, with one of his own 3½ft achromatic refractors;

Edward Nairne, instrument maker, with one of his own 2ft Gregorian reflectors.

Although all these observations may be classed as successful, results were somewhat disappointing, the recorded time of observation of the first external contact of the planet on the limb of the Sun varying as much as 36 seconds between the seven observers, a fact which Maskelyne ascribed to uncertain atmospheric refraction due to the low altitude. Fig. 4.2 shows the appearance of the 'black drop' as seen by Hirst (see p. 34 above).[12]

Worldwide, 150 successful observations were reported from 77 different locations.[13] How many observers were clouded out we do not know.

The Board of Longitude and The Nautical Almanac

Most of the Board's business in this period concerned with the Harrison affair and with the Nautical Almanac has already been told in Chapters 8 and 9. The Nautical Almanac for 1768 was published on December 30 1767, for 1769 in October 1768, for 1770 in October 1769, and for 1771 in January 1770 (eleven months ahead), one thousand copies of each being printed, selling at 2s.6d.[14]

Between 1766 and 1768, the Board awarded the instrument maker

John Bird a total of £650 for writing illustrated descriptions of his method of dividing astronomical instruments and of his 8ft Greenwich mural quadrant, a half-scale model of which was deposited in the British Museum.[15] The publication of these pamphlets by the Board at one shilling each, the first in April 1767[16], the second in July 1768, further stimulated the demand for mural quadrants of this design for observatories at home and abroad, by Bird himself, by Jonathan and Jeremiah Sisson, and by Jesse Ramsden.[17] The present whereabouts of the model sent to the British Museum is not known.

Bradley's observations

The affair of the dispute on the ownership of Bradley's original observation results, last discussed on p. 55 – which Maskelyne and astronomers the world over wanted to be published for the public good – had been prosecuted originally by the President and Council of the Royal Society in their capacity as permanent Visitors to the Royal Observatory. After Bradley died in 1762, his books and astronomical papers were left at the observatory by his executors, his successor Bliss giving his receipt for them. Nonetheless, the Council had, in response to Maskelyne's memorial at the Society's meeting of 8 June 1763, taken advice as to whether the Royal Society had a legal right to the possession of such papers. In May 1764, the Attorney and Solicitor Generals gave their opinion that the President and Council did *not* have such a right because Queen Anne's warrant of 1710 appointing them as Visitors was no longer applicable. But, they said, the Crown might have such a right. It was at this stage that steps were taken to ensure that such a situation could not arise again, as described in Chapter 6.

Bliss died on 3 September 1764. Eight days later, William Dallaway, one of Bradley's executors, came to the observatory and demanded from Mrs Bliss all Bradley's books and papers, producing her late husband's receipt.[18] The following letter from Dallaway to the Society's Secretary explains the executors' attitude:

The Executors of Doctr. Bradley (in whose discretional Power all his Manuscript Books and papers were left) always intended to compliment the Royal Society (if they would accept of them) with all the Doctrs. Books of Observations, together with whatever other papers cou'd be found of his, that might be of use to the Publick.

This compliment had long since been made, had not some forwardly Members of that Society drop'd Reflections Injurious to the Doctors Character; and endeavour'd likewise to obtain by Compulsion, what they had not the least right to.

Wm. Dallaway

Brimscomb   Oct 22d 1764[19]

In view of the law officers' opinion that the legal initiative must now lie with the Crown, action shifted from the Royal Society to the Board of Longitude (an official government body) – though the change of influence was not as great as might be thought because the two individuals most concerned, the President of the Royal Society and the Astronomer Royal, were both Commissioners of Longitude *ex officio*.

At the Board meeting in June 1765, Maskelyne said he understood that the executors were ready to deliver up the books 'to such persons as were ready to receive them', and the chairman, Lord Egmont, undertook so to inform one of the Secretaries of State. However, there were still legal complications and it was decided to wait until Bradley's only daughter came of age in January 1767. Then, at their meeting on 12 December 1767, the Commissioners were somewhat mortified to be informed officially about a letter of the previous 18 May from Miss Bradley to Lord Shelburne, Secretary of State, in which she said that, as no one on behalf of the King, the Board, or the Society had asked for them, she had, on coming of age, given the documents to her uncle, the Rev. Samuel Peach, and that in any case eminent Counsel concurred in the opinion that they were her sole property.[20] Actually, this came as no surprise, as Thomas Hornsby, Savilian Professor of Astronomy at Oxford and one of the Commissioners, had kept in touch with the family (who lived in Oxford), and Dallaway had written to the Royal Society to say what had happened.[21]

The Board once more took legal advice, only resulting eighteen months later in the letter from Miss Bradley and Mr Peach absolutely refusing to give up the documents 'without a very valuable consideration'.[22] Meanwhile, Lord Morton, one of the principal protagonists, had died, to be succeeded in November 1768 as President of the Royal Society by James West. Preparations were once again made to take the matter to court on behalf of the Crown. And there – such is the speed of the Law – matters rested until 1772.

Meanwhile, Maskelyne had received Halley's papers, bought by the Board of Longitude from his daughter, Mrs Catharine Price, for £100 in 1765,[23] and Bliss's Greenwich records were still in the observatory. Soon, in 1771, the Board of Longitude were to pay £100 to Mrs Elizabeth Tew of Islington for Flamsteed's books and papers.[24] But Flamsteed's observations had already been published, Halley's left much to be desired in accuracy – 'he [recorded] every observation possible, even bad ones, to fill up the Saros the better'[25] – and Bliss's covered only two years: these were no compensation for the lack of Bradley's observations, whose publication was so badly needed, not only by Maskelyne to test the validity of the predictions in the Nautical Almanac, but also by astronomers and mathematicians abroad such as Lalande and Laplace, working on new theories in celestial mechanics, so important to navigators and astronomers.

The Royal Society and the Mason–Dixon Line

Charles Mason and Jeremiah Dixon had returned from St Helena early in 1762. In August 1763, they signed an agreement with Thomas Penn and Lord Baltimore, the hereditary proprietors of the provinces of Pennsylvania and Maryland, to go to North America to help the local surveyors define the disputed boundary between the two provinces. Arriving in Philadelphia on 15 November 1763, they disembarked with their instruments, which included a transit instrument and a 6ft zenith sector by Bird, similar to that used by Maskelyne in St Helena but with an improved plumb-line suspension. They began their operations before Christmas 1763.

At a meeting of the Royal Society's Council in London on 24 October 1765, the Astronomer Royal – not yet a Council member but coopted for the occasion – said he had received a proposal from Mason and Dixon that, when they had completed the surveying work for the Proprietors, they might spend some time furthering the cause of science generally – and of the Figure of the Earth in particular – by making a precise measurement of a degree of latitude in Maryland, to add to the data provided by similar measurements by Picard and the Cassinis in France before 1718, by Bouguer and La Condamine in Peru and Maupertuis in Lapland in 1735–6, by Maire and Boscovich in the Papal States in 1750, and by Lacaille at the Cape of Good Hope in 1752.

What a splendid opportunity, said Maskelyne. The Council agreed that such an operation appeared:

> . . . to be a work of great use, and importance: and, that the known abilities of Messres Mason and Dixon, the Excellence of the Instruments with which they are furnished, the favourable level of the Country, and their having assistants well practiced in Measuring; [and they] do all concur in giving Ground to hope, that this business may soon be executed with greater precision, than had ever yet

been done; and at much less charge than the Society can expect an opportunity of doing it hereafter.[26]

The Council resolved that the two observers should receive £200 for this extra task and directed Maskelyne to send technical instructions and additional instruments – four 10ft fir measuring rods, a 5ft brass standard scale, a 6ft spirit level, three thermometers, and the Society's clock by Shelton, the same that Maskelyne had taken to St Helena and Barbados, all despatched in seven cases in the *Ellis* merchantman, Captain Samuel Egdon, from London to Philadelphia on 6 December 1765.[27] Thomas Penn and Lord Baltimore gave permission for their instruments to be used in the new project.

Then, in June 1766, the Council was horrified to hear that the *Ellis* had been wrecked and that there was a report that their precious clock had been damaged. Nothing had been heard from Mason and Dixon since the previous year so a letter was sent asking for the clock to be returned so that it could be repaired in time for the transit of Venus in three years' time.

The letters had indeed gone astray and Mason and Dixon did not receive Maskelyne's instructions nor the Proprietors' permission until 1 October. However, they then began operations almost immediately – laying out a line running due north for some 80 nautical miles from the south-west corner of Delaware in what is today called the Delmarva Peninsula, east of Chesapeake Bay. In fact, the only damage the Society's instruments received in the wreck was a broken pendulum suspension spring, so they ignored the order to return the clock. Between December 1766 and March 1767, in very cold weather, they made their first measurement of the arc of the meridian, using the transit instrument and clock to give direction, and the zenith sector to find the precise latitudes of the two extremities. They also made gravity measurements, similar to earlier ones made with the same clock in St Helena, the Cape of Good Hope, and Barbados.

They packed up the clock in June 1767, leaving it in Philadelphia to be sent home: by November, it had been set up and was going at Greenwich. Having finished their operations for the Proprietors by the end of January 1768, they spent the months until mid-June remeasuring the Society's meridian arc. On 11 September, 1768, they embarked in the *Halifax* packet boat at New York for passage to Falmouth and a few months' rest in England before undertaking their next expeditions. 'Thus ends my restless progress in America', wrote Mason in his journal.[28]

In the meantime, considerable thought was being given in England to the preparations needed for the 1769 transit of Venus. Maskelyne, elected with Dr Benjamin Franklin to the Royal Society Council on St Andrew's Day, 1766, was a member of a special committee set up to consider where British expeditions should be sent, what equipment was needed, and who should be the observers; it was he who summarized papers by Short, Bevis, Ferguson and himself on 30 November 1767.

The King granted £4000 to fund the expeditions and the Admiralty promised ships to go to the North Cape of Norway and to the South Sea.

Though Maskelyne played a most important part in their scientific planning and analysis, space permits no more than a brief summary of the British expeditions, which are listed in the order in which they left England. The transit occurred on 3 June 1769.

> *Hudson's Bay* (observers William Wales [who had originally said he would only go to a warm climate] and Joseph Dymond): they left Yarmouth in a Hudson's Bay Company ship 23 June 1768, reaching Fort Churchill 8 August: spending the winter there, they successfully observed the transit and sailed for England 7 September 1769, reaching Plymouth 16 October.
>
> *Tahiti, South Pacific* (Lieutenant James Cook, RN, and Charles Green): the *Endeavour* bark, purchased for the expedition, sailed from Plymouth 26 August 1768, with naturalists Joseph Banks, Daniel Solander and party embarked, reaching Tahiti via Cape Horn 13 April 1769: Cook and Green observed the transit at Fort Venus, Matavai Bay, other officers and gentleman observing at other sites in the island, all successfully: the overt purpose of the voyage accomplished, Cook left Tahiti 13 July 1769 to carry out the secret part of his orders – to search for an imagined continent south of Tahiti and westwards to New Zealand, and to chart the coasts of New Zealand: the Great Southern Continent was not found, but New Zealand was charted and, on the way home, the east coast of Australia discovered, charted and claimed for King George: Green died of fever 29 January 1771 soon after the ship left Batavia ('he had long been in a bad state of health, which he took no care to repair, but on the contrary lived in such a manner as greatly promoted the disorders he had had long upon him, this brought on the Flux which put a period to his life'[29]): arrived the Downs 13 July 1771.
>
> *Cavan, near Strabane, Co. Donegal* (Charles Mason): reached

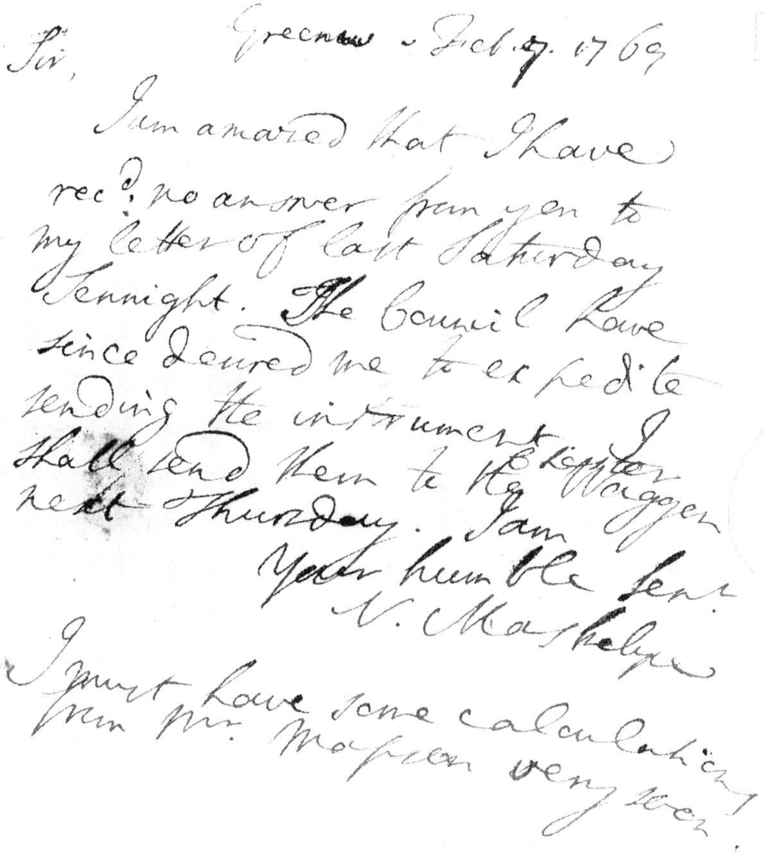

Sir, Greenw. Feb. 7. 1769

I am amazed that I have recd. no answer from you to my letter of last Saturday sennight. The Council have since desired me to expedite sending the instruments. I shall send them to the Waggon next Thursday. I am

Your humble Servt.

N. Maskelyne

I must have some calculations from Mr. Mapson very soon.

Fig. 10.3. An Example of Nevil Maskelyne's Handwriting. This letter to Charles Mason, dated 7 February 1769, concerns the arrangements for the latter to go to Ireland to observe the transit of Venus on 3 June. Mason reached Londonderry in March. John Mapson, mentioned in the postscript, was one of Maskelyne's almanac computers. (RGO MS 4/184:4.) (Photo: RGO)

Londonderry March 1769: transit observations successful: other observations until 28 November 1769.
Northern Norway (William Bayly and Jeremiah Dixon): sailed in the *Emerald* 32 from Sheerness 13 April 1769: as an insurance against bad weather on the day, they set up observatories about sixty miles apart, Bayly at North Cape, Dixon on Hammerfest Island: observations successful: arrived Sheerness 30 July 1769.
The Lizard, Cornwall (John Bradley): something of an after-

thought suggested by Maskelyne, to use the transit observations to settle the latitude and longitude of the Lizard, the point of landfall and departure for so many voyages to and from the English Channel: Bradley and Mr Hunt, master of the *Arrogant* 74, sailed together from Portsmouth for Falmouth in the *Seaford* 22, 11 May 1769, Bradley returning in the *Cruizer* sloop 4 July 1769.

All but the last were funded by the Royal Society from the royal grant; the Lizard expedition was funded by the Board of Longitude. The main instruments provided were the same for each expedition (two sets for Norway): one or more 2ft Gregorian reflectors by Short or Bird, a 12in astronomical quadrant by Bird, and a clock by Shelton (by Ellicott for Hudson's Bay). Each also had a portable observatory. The instruments bought in 1761 were all used, augmented by new ones. Because Short had died in 1768, Bird provided the new Gregorians.

Maskelyne provided formal instructions for each observer and also published a fifty-page 'Instructions Relative to the Observation of the ensuing Transit of the Planet Venus over the Sun's Disk, on 3rd of June 1769' as an appendix to the Nautical Almanac for 1769.

11

The 1770s

Personal and family

Maskelyne was a methodical man and, luckily for us, his descendants were caring people who had a proper sense of history. Three account books in his own hand have survived, recording every penny he received and every penny he spent, of both public and private money, between August 1773 and his death thirty-seven and a half years later. In addition, there are six pocket 'Memorandum Books' and two 'Travelling Books', covering the period from 1776 to 1806, which form a sort of informal diary, interspersed with addresses, recipes, medical prescriptions, travel details (sometimes even cost per mile), what wine was laid down, what ales and elixirs were made, what game was received, who came to dinner, anecdotes, and even jokes.

These are fascinating documents which, together with the many letters that have survived, tell us a great deal about the social conditions of the time as well as about Maskelyne's own life. A few extracts are given at the end of this and subsequent chapters.

In 1768, soon after returning from India, brother Edmund had accompanied Robert and Margaret Clive on a continental tour, afterwards settling down to enjoy his new estate at Basset Down. In December 1771, he married Mrs Catharine Muscott, late of Ludlow – both second marriages.

The 1770s were years which brought much sadness to the Maskelynes. In 1772, Nevil's eldest brother, the Reverend William, living at the Ponds, Purton Stoke, died at the early age of 47. He died intestate so his elder brother Edmund inherited. But there was a complication: in 1748, before William took orders, Jane Secker (as she became known) was born; her putative father was either William Maskelyne or his first cousin John Walsh, but William took responsibility – and on his death this responsibility fell first upon Edmund, then upon Nevil. About 1777, she married William Sacheverell, who was in business in Oxford Street,

London. If she had been legitimate, she would presumably have taken the Maskelyne inheritance.[1]

Then, on 25 April 1774, Uncle Nevil Maskelyne, of the Down, Purton, died at the age of 82. He had fallen upon hard times and had little to leave except the house. Meanwhile, the affairs of both Uncle James Houblon Maskelyne and brother Edmund were causing worry for Nevil and Margaret. James had a wife and two daughters, but no marriage lines and no money.[2] As for Edmund, first, his health was deteriorating: 'For God's sake,' he wrote in March 1773 in a letter to Nevil largely concerned with the setting up of a fund to relieve Uncle James's difficult situation, 'take care of your health & do not study too intensely. I have shattered my Nerves not a little by too much Business.'[3] Then, a few months after Uncle Nevil's death, both Nevil and Margaret had long letters from Uncle James alleging that Edmund's new wife, in concert with Blanche Floyer, (his first wife's mother, or perhaps aunt), was scheming against the Maskelyne family, helped by an error in the drafting of Uncle Nevil's will. 'I have long suspected Mrs Floyer, to have had views and designs upon the Captain's Fortune, ever since she has known him, and that ever since his marriage with his present Wife, his Wife and Mrs Floyer have acted in Concert in carrying on Schemes, to get for their own Family, from You and the Captain's own Relations his Estates and Fortune; I say I suspect I don't say 'tis so. The Day after I came to Basset Down came Mr Muscott Mrs Maskelyne's Son, Lord perhaps of Purton, and perhaps of the Maskelyne Inheritance, which the Family has enjoyed 700 years. . .'[4]

Subsequent events proved that these allegations had much substance. In the end, as we shall see, the family succeeding in circumventing this scheming as far as the estates were concerned, though not the fortune, which, thanks to Edmund's Indian days, was not inconsiderable. But meanwhile, a much more immediate tragedy supervened. On 22 November 1774, racked with pain from illness which he thought was incurable, Robert Clive died, some say by his own hand. His widow, Margaret, Lady Clive, retired to Oakly Park, near Ludlow in Shropshire, where she lived until her death in 1817, and whither Nevil was to make frequent visits.

But family troubles seldom come singly. The following August, Nevil was summoned to Basset Down where Edmund lay dying. Margaret also visited him before he died – on 14 September 1775, aged 47. By his will, the two estates at Purton – the Down and the Ponds – were left to Nevil, Basset Down to Catharine his widow for life, then to cousin Thomas Carter of Foxley (executor), then to Nevil. The residue, quite a considerable sum, went to his widow.

A few months before Clive died, the Rectory of Shrawardine in the Diocese of Hereford and the County of Salop, which was in Clive's gift, came vacant and his son, now the second Lord Clive, offered it to his uncle. Having obtained the King's permission to accept, Maskelyne was inducted on 3 February 1775. He remained Rector there until 1782, visiting once a year or so, the parish being looked after from day to day by the Rev. Thomas Hanmer. This brought him tithes of £140 annually. Until his marriage, he also had an income from his Fellowship of Trinity College, Cambridge, amounting to about £110 annually.[5]

The degree of Doctor of Divinity was open to Bachelors of Divinity of at least five years' standing and Maskelyne put his name forward in 1777 after nine years as a BD. To qualify, he had to keep an Act against a doctor and preach a Latin sermon in the University Church. The degree was considered the highest intellectual honour the University could bestow and the candidates were expected to attain a standard of scholarship worthy of it.[6] He records in some detail in his Account Book the immediate cost to himself in taking that degree:

1777

June 9–17	Journey to Cambridge & back, & expences of Doctor of Divinity's degree.	
	Journey to Cambridge & back June 9	
	Postchaise 54 miles 2£.0S.6D Postboy 5S. Turnpike 4S. Dinner 7S.	2.16. 6
	Do. back June 16 in fly	0.17. 0
	My expences at Cambridge	0.17. 2
	Servants expences on journey & at Cambridge	1.10. 6
	Proctors Man	0. 2. 6
	Notary Public	0.12. 6
	Bursar of Trinity College	25.10. 0
	Dr. Smith's man (who lent me the Dr's robes)	0. 5. 0
	Bluecoat doorkeeper to the Schools	0. 7. 0
	Vice Chancellor	4. 0. 0
	Bell-ringer	0. 2. 0
	Harry Gordon, Butler's bill	4.11. 6
	Proctor compounding &c	10.16. 8
	Register	2. 3. 0
	Father	1. 7. 0
	Professor	1. 7. 0
	University Servants fees	1. 1. 0
	Professor's Servant	0. 5. 0

Dr. Smith's man who lent me Dr's robes		0. 5. 0	
College Porter		0. 2. 0	
Mr. Day writing Caution Grace & Favour		0. 2. 0	52.19. 0
June 6	Pd. Mr Theed in Wich Street for a new Gristle wig		2. 2. 0
25	Pd. Mr Aswell of Greenwich for a new hat		0.18. 0
26	Pd. Mr Jno. Brewerton Robe Maker		
	A double Alepeen Gown & Cassock	6.19.10	
	Silk Sash	0. 8. 0	
	Dr. of Divinity's Scarlet Hood	2.11. 3	
	Box & lock & key	0. 8. 6	10. 5. 7
June 27	to July 3. Journey to Cambridge & back		
	Fly to Cambridge June 27	0.17. 0	
	My Expences at Cambridge	1.16. 2	
	Fly back to London	0.17. 0	
	Servants expences on journey & back & at Cambridge	2. 0.11	5.11. 1

Total of Expences of degree & 2 journeys to Cambridge
64£.5S add Gown & Cassock wig & hat 13£.5.6 makes 77£.11S.[7]

The observatory

In 1770, Maskelyne complained to the Visitors of the inadequacies of Graham's old equatorial sector used for observing comets: because the telescope was of 'the old sort', faint comets were difficult to see; because it had to be used in the Great Room, it had to be moved from window to window, high comets could not be seen, and it lacked stability. He suggested two new rooms be built, to the east and west of the Great Room. A single new equatorial sector would suffice, he said, because it could be moved from room to room according to the position of the comet.

The Visitors in their wisdom, decided it would be better to convert Wren's summer houses by removing the pineapple roofs, adding a storey, and fitting 'conical movable roofs'. This was done in 1773, and two equatorial sectors were provided by Jeremiah Sisson and two clocks by John Arnold.

How wrong the Visitors were! Of these new observatories, Airy had this to say in 1855:

The positions of these two instruments are so strangely misadapted to their purpose, that I am utterly at a loss to conceive under what circumstances their places were selected. The Octagon Room towers over them in such a manner that nearly the whole sky from south to east, to the height of 40° or more, is hidden

Fig. 11.1. *Nevil Maskelyne*, 1779, aged 47, by John Downman. From an oil painting signed and dated *Jno. Downman Pinxt. 1779*, in the possession of the National Maritime Museum. There is a slight doubt whether the sitter is Nevil Maskelyne, but it could be the first portrait referred to by Lady Clive in a letter to Margaret Maskelyne dated 30 October 1813: 'The portrait I received from my Brother, is not so like him as that I send, in his canonical dress. . .' The second portrait referred to is presumably that by Van der Puyl in 1785 (Fig. 13.1). (Photo: NMM)

from the North-west Dome; and nearly the whole sky from south to west, to the height of more than 50° in some parts, is hidden from the North-east Dome.[8]

Airy converted the North-west dome into two bedrooms in 1856: the author and his secretary found them most pleasant offices from 1969 to 1982. As for the sectors, Maskelyne condemned them both as useless in 1774. They had basic design faults and, despite modifications, never proved satisfactory.

Other changes in observatory equipment were more successful, however. In 1772, Dollond provided achromatic object glasses for the transit instrument and for the South Quadrant, making it much easier to observe faint objects.[9] In 1774, Dollond sold the observatory a 46in triple achromatic telescope with an aperture of 3.6in, for £63 (Fig. 7.4), to take the place of the instrument he had lent in 1765. This was by far the most successful small telescope the Royal Observatory ever had, being used regularly until 1838, 1084 observations of eclipses, occultations, etc., being recorded in those sixty-four years.

In 1778, at his own expense, Maskelyne installed a camera obscura in the western turret room of Flamsteed House.

In 1779, to keep the rooms cool, extensive alterations were made to the New Observatory: the roof shutter openings in the Transit and Quadrant Rooms were increased from six inches to three feet, and the flat ceilings were removed and sloping double roofs substituted. The movable quadrant observatory had its circular roof replaced by a sloping roof with sliding shutters, not unlike a lean-to greenhouse: henceforth known as the Advanced Room, it was used instead of the Great Room for observations of occultations, etc., by small telescopes. The movable quadrant was abandoned.

Turning to personnel matters, Maskelyne submitted a Memorial to the King in April 1771, praying for a pension for his Assistant of at least £70 per annum, to augment the £26 salary granted by Charles II in 1675. It was a responsible job, he said, and his last Assistant had left simply because the pay was so low, now that the 1765 regulations forbade the Assistant – quite rightly – showing strangers around the observatory and pocketing a fee. The prayer was duly granted and the Assistant then received £86 (£10 went in taxes),[10] remaining so until 1810 despite considerable inflation in that period. Of course, his lodging was free and he ate at the Astronomer Royal's table.

Reuben Burrow replaced William Bayly as Assistant on 25 March 1771. The latter had been at Greenwich for just over four years, with a short break to go to the North Cape for the transit of Venus. He did not have long to wait for a new job, being recommended by Maskelyne in December of the same year to go with Cook on his second voyage (of which more anon). Burrow did not last long, being replaced by John Hellins at Michaelmas 1773. As will be discussed in Chapter 12, Maskelyne left the observatory in Hellins's care for six months of 1774 during the Schiehallion expedition (where he had Burrow as assistant), but was not best pleased with what he found on return. Writing in 1790 to Trinity College, Dublin, where Hellins was a candidate for the post of Professor of Astronomy, Maskelyne said that, though capable of the common business of observing, Hellins was 'the least serviceable Assistant I ever had, especially in the calculation of observations, in which he made so little progress that I thought it necessary to part with him.'[11] This parting came in March 1776, when George Gilpin arrived, to remain more than five years.

Foreign contacts continued. John Ewing came to England in 1773 to solicit funds for an academy in Delaware and the observatory in Philadelphia already mentioned. However, the news of the Boston Tea Party of December 1773 did nothing to attract English support. 'In the present unhappy situation of American affairs,' wrote Maskelyne to Ewing in August 1775, 'I have not the least idea that anything can be done towards erecting an Observatory in Philadelphia, and therefore cannot think it proper for me to take any part in any memorial you may think proper to lay before my Lord North at present.'[12] No observatory was established in Philadelphia then or even after 1776.

England's Royal Observatory was looked upon as the model for other observatories being erected all over Europe at this time, and it was English instrument and clock makers that supplied most of the equipment. In the 1770s, Maskelyne was often consulted and himself acted as the go-between for the Bishop of Agria and the Elector Palatine for equipping their respective observatories in Eger in Hungary and in Mannheim in Germany, with instruments from Sisson and Arnold.[13] In the autumn of 1777, Thomas Bugge, Astronomer Royal to the King of Denmark, toured England, visiting Cambridge, Oxford (where he visited Shepherd in the new Radcliffe Observatory), and of course Greenwich. He visited private observatories and instrument and clock makers in London, making sketches of many of the things he saw, though they were never published.[14]

Maskelyne was elected Fellow of the Göttingen Royal Society in 1771 and of the St Petersburg Academy in 1776.

In the *Philosophical Transactions* of 1772 and 1777, he published details of two inventions: modifications to the Hadley sextant, and a prismatic micrometer for achromatic telescopes. Neither came into general use.

The Board of Longitude

The Board generally met at the Admiralty under the chairmanship of the First Lord of the Admiralty (the Earl of Sandwich in the 1770s) – or if he was absent, the Speaker of the House of Commons – on Saturdays three times a year, in March, June and November, with occasional extra meetings if needed. We will discuss first the Board's publications, then what today would be termed its hardware.

The 1770s were for Maskelyne a period of consolidation. Aged 38 at the beginning of the decade, he had the computation and publication of the Nautical Almanac running very smoothly, publishing one year ahead in 1770, four years ahead by 1780. At 3s 6d unbound, sales ran at about 800 a year. Although the almanac reduced the time of working out each lunar distance sight from four hours to about thirty minutes, correcting the observed distance for the effects of parallax and refraction was still very laborious. The ultimate tables were those of Shepherd, published by the Board in 1772.[15] However, as they weighed more than 5 kg and were some 90 mm thick, one wonders how much they were used at sea. Luckily, Maskelyne published several alternative tables of much smaller bulk, if less convenient to use.

When he returned from his Irish expedition in 1770, Mason set to work comparing Mayer's tables with Bradley's actual observations. The revised lunar tables he produced were used in all almanacs published after 1774. Mayer's original tables had eventually been printed in 1770, with a preface by Maskelyne in both English and Latin.[16]

They say that imitation is the sincerest form of flattery. In 1772, the *Académie de Marine* in Paris arranged for the publication in Brest of lunar-distance tables based on the Greenwich meridian for the last eight months of 1772 and for 1773 – with the instructions translated into French – copied (by permission, one assumes) from the *Nautical Almanac*. That same year, 1772, Lalande published similar tables and explanations in the *Connaissance des Temps pour l'Année 1774*, based on the Greenwich meridian despite the fact that all other tables in the almanac were based on the Paris meridian. 'I believe in rendering service', he wrote to the

Académie in 1775, 'by inserting these important calculations that we have received thanks to the zeal and munificence of the Government of England and the Board of Longitude. Maskelyne, with scholarly zeal, sent them and will send future ones as soon as possible.'[17]

This cooperation between Maskelyne on the one hand, and Lalande and his editorial successors Jeaurat and Méchain on the other, continued despite France's position on the opposite sides to England in the American and subsequent wars from 1778. Maskelyne always managed to get his tables to Paris in time to be translated for the *Connaissance*. Then, in the *Connaissance* for 1789 (published in 1786), Jeaurat published the English tables, but converted to Paris time. The *Connaissance* for 1790, however, (published in 1788) had tables for Paris time calculated entirely in France by M. l'Emery at the expense of the Maréchal de Castries. Happily, as we shall see, this fruitful Anglo-French cooperation was to continue, wars and revolutions notwithstanding.

All of this, together with trials of watches and instruments, etc., made a great deal of extra work for Maskelyne. To compensate him, the King authorized him to claim, from December 1774, £15 for each Board of Longitude meeting attended, the money to come from the Navy Board's 'fund of old naval stores', a privilege which the university professors on the Board had had since 1762.

In 1771, the Board received the names of fourteen masters of the Royal Navy who had, as requested in 1768, been instructed in lunar observations by George Witchell and John Bradley, respectively Headmaster and Mathematical Usher at the Royal Academy at Portsmouth. At the next meeting, however, they heard that some of the older masters refused even to try to learn how to use these new-fangled methods. The Board backed down, recommending that their Lordships change the rules so that the need for certificates of competence in lunar observation should apply only to future appointments of masters.[18]

Bradley's observations

As for the recovery of Bradley's observations (see p. 107), matters dragged on slowly, a law suit having been started. Then in 1776, just as the matter was about to come to court, the Reverend Samuel Peach junior (whose father originally had the papers but died in 1769) married his cousin Miss Bradley and became the legal owner of the observation books, which he forthwith presented to Lord North, Prime Minister and also Chancellor of Oxford University. Despite a deputation from the

Board to ask him to deposit them at the Royal Observatory so that they could be published at public expense – the first ten years of Maskelyne's observations were so published the same year – the Prime Minister insisted they should go to Oxford for publication there, edited by the Savilian Professor of Astronomy, Thomas Hornsby – who was a Commissioner of Longitude. The law suit was dropped.

Cook's second voyage

Although the term *chronometer* did not come into general use until the mid-1780s, it will for convenience be used in this book from now on, to describe timekeepers designed for finding longitude at sea, despite the fact that their users may have employed other terms.

We last met the watchmaker Larcum Kendall in 1767 when Maskelyne sent Harrison's H4 to Kendall's workshop so that he could copy it (p. 84), for doing which the Board of Longitude agreed to pay £400, half to be paid immediately, half on completion. At the Board meeting of 13 January 1770, Kendall presented his chronometer, which we will henceforward call K1. William Harrison, who was attending, pronounced it a satisfactory copy of his father's masterpiece. The Board were impressed and granted Kendall an extra £50, both H4 and K1 being given to Maskelyne to try together at Greenwich. Then, at their May meeting, Kendall offered to make a chronometer just as good for half the price, an offer which the Board accepted. At the same meeting, the Board met for the first time the 34-year-old watchmaker John Arnold, who produced a chronometer which he said could be made for only 60 guineas. To continue his experiments, the Board advanced him a total of £500 in 1771.

On 13 July 1771, the *Endeavour* bark, Lieutenant James Cook, anchored in the Downs after her epoch-making voyage to the South Sea. Alas, Charles Green was not among those who returned, but Cook and Maskelyne between them assembled his somewhat garbled transit-of-Venus records and submitted them, with Cook's own, for publication.[19] So successful had that voyage been that the Admiralty decided in September that Cook – by now promoted to Captain – should lead another. He had already proved that there was no Great Southern Continent in temperate latitudes in the south-west Pacific: now he was to see whether it existed anywhere else. On 25 October, Maskelyne wrote to Lord Sandwich:

representing that the intended expedition to the South Seas may be rendered more serviceable to the improvement of Geography & Navigation than it can otherwise be if the ship be furnished with such Astronomical Instruments as this Board hath the disposal of or can obtain the use of from the Royal Society and also some of the Longitude Watches; and, above all, if a proper person could be sent out to make use of those Instruments & teach the Officers on board the ship the method of finding the Longitude.[20]

So, to a primary geographical aim, was added a secondary technical aim – to try out whether chronometers were a practicable and useful means of finding longitude at sea.

On 28 November, the Board, who had been informed that Cook was to have two ships instead of one, endorsed Maskelyne's proposals, agreed to pay to send astronomers and their instruments in each ship, and decided that K1 should go too. Arnold was asked to have four chronometers ready by the end of January. John Harrison had completed another chronometer, H5, and the Board proposed sending this too, but his son refused, saying 'the trial would take up too much time'.[21] (John was now 78.)

A fortnight later, Maskelyne proposed two names to go as the expedition astronomers: William Wales, one of his Nautical Almanac computers and Charles Green's brother-in-law; and William Bayly, until a few months before his assistant at Greenwich. They agreed to go for £400 per year inclusive of expenses. It was planned that, whenever the ships stayed in any harbour for more than a few days, the astronomers should land, set up their portable observatories to protect the instruments from the elements and the natives, and set the pendulum clocks going. Then they would make frequent observations for geographical position – latitude by meridian altitude with the 1ft astronomical quadrant, longitude by lunar distance (as at sea), and by Jupiter's satellite eclipses if possible, all timed by the clock – and observations to measure the rate of going of the clocks and chronometers. Maskelyne produced lists of the instruments needed, most of which had been bought for the various transit of Venus expeditions, and drafted technical instructions for the observers.

Although there were some disagreements over Green's observations, relations between Cook and Maskelyne were good, the former dining as the latter's guest at the Mitre Club. In September 1775, Maskelyne gave a dinner party at Greenwich for Cook, attended by Pringle, the Royal Society's President, General Roy, and Daniel Solander, Cook's naturalist in his first voyage.[22]

The *Resolution*, Captain James Cook, and *Adventure*, Captain Tobias Furneaux, sailed from Plymouth on 13 July 1772, Wales with K1 and Arnold's box chronometer No. 3 in Cook's ship, Bayly with Arnold's Nos. 1 and 2 in Furneaux's. As the award of considerable sums of public money depended upon the result of the trials, each ship's chronometers were kept in boxes each of which had three keys, one kept by the Captain, one by the First Lieutenant, one by the Astronomer.

Meanwhile, Harrison had enlisted the sympathy and help of no less a person than the King. Early in 1772, his new chronometer, H5, was put on trial at the King's private observatory at Richmond. At a Board meeting on 28 November, he presented a memorial stating that during the ten-week trial H5 had lost only 4½ seconds and was thus within the limits prescribed by the 1714 Longitude Act: would the Board please give him a certificate so he could collect the second £10 000 specified in the 1765 Act? (see p. 79) No, they told William Harrison, they 'did not think fit to make any Alteration in the mode they had already fixt upon for making the trial of his Father's Time-keepers. And no regard will be shewn to the result of any Trial of them made in any other way.'[23]

In August, William Harrison had written to the Prime Minister, Lord North, reciting his grievances, but, having had no reply – and having heard the Board's November verdict – he wrote a similar letter in December to John Robinson, Secretary to the Treasury. In reciting the allegedly unfair conditions imposed by the Board in 1765 before the second £10 000 would be granted – that two more chonometers must be made by the ageing watchmaker, and that there must be a twelve-month trial under unspecified conditions – he concluded with the following grievance which is relevant in the context of this book:

And they require ten months of the trial to be under the entire Power and Management of Mr Maskelyne, who, by every Tie of Interest and a most decided part for many years taken against us, stands pledged to crush the Invention.[24]

On 2 April 1773, Harrison presented a petition to Parliament, asking for justice, laying the blame squarely upon the Board of Longitude, and asking Parliament to grant him the second £10 000 forthwith. On 27 April there was a lively debate in the House of Commons in which sympathy or otherwise for the petitioner seems to have depended at least partly on party lines. Following Lord North's suggestion that the House should hear the Board's side of the story, the Whig leader Edmund Burke made an impassioned speech on Harrison's behalf:

Mr. Speaker: I must own, I am astonished by the replies which have now come from the other side of the House. Where, Sir, is the dignity, where is the sense, where even the justice of the representative of a great, powerful, enlightened, and maritime nation, when a petition is laid before them, claiming not favour, but justice; claiming that reward which law would give him, and to see it refused

– upon what principle? Why, a man of 83 [actually, he had just turned 80] is to make new watches; and he is not only to make them, but to make new voyages to the Indies to try them. Good God, Sir, can this be a British House of Commons? This most ingenious and able mechanic, who has spent above 40 years of an industrious and valuable life in search of this great discovery – and has discovered it; who has, according to the verdict of the whole mechanical world, done more than ever was expected – and even gone beyond the line drawn by the Act of parliament, in consequence of which he undertook the work – this man is now, at the age of 83, to have his legal right withheld from him, by adhering to an Act which stands between him and his reward. . . . Is this a conduct worthy of the munificence of this House? Of a nation that owes to her navigation, her wealth, her consequence, her fame![25]

Eloquence notwithstanding, the Harrisons' parliamentary friends advised that, on legal grounds – and that was what the House had to debate as a result of the petition, not the emotional aspects – the Board of Longitude had acted entirely properly according to the duties imposed upon it by the Act of 1765. On 6 May, therefore, Harrison withdrew his original petition and substituted another, based this time not upon the Act of 1765 but, it could be said, upon emotion – resting his pretensions on the Bounty of Parliament and 'praying for such assistance as to the House shall seem meet'. This new petition was presented to the Commons on 11 June by Lord North, who said that the King had seen it and recommended it to the House. The Act 13 George III c.77 received the Royal Assent on 1 July 1773, granting Harrison a sum not exceeding £8750 for 'having . . . applied himself, with unremitting Industry for the Space of Forty-eight Years, to the making of an Instrument for ascertaining the Longitude at Sea . . . as a further Reward & Encouragement over & above the Sum already received by him for his Invention of a Timekeeper, and his Discovery of the Principles upon which the same was constructed.'[26]

He had not, after all, complied with the provisions of the Act 5 Geo. III c.20, for which he had to pay a penalty of £1250. As to Maskelyne's attitude to the whole affair, I cannot do better than to quote the words of the late Professor Eric Forbes in an article in which he suggested that the title 'The man who found longitude' could equally well be conferred upon Tobias Mayer:

Popular books faithfully repeat the traditional story of Harrison as a poor illiterate Yorkshire carpenter who suffered at the hands of the Board to such an extent that he finally had to seek the personal intervention of King George III to obtain his just reward. Colonel Quill's biography, however, attempts to correct the exaggerations and distortions which have grown with that story, due primarily to Harrison's attacks on the Board and its individual members. The Hannover correspondence [to and from Mayer] lends strong confirmation to the circumstantial evidence cited by Quill. There is no truth in Harrison's accusation that the Board's decisions were influenced by any prejudice against timekeepers, since Mayer's claim was treated even more harshly than Harrison's.[27]

Maskelyne himself, writing in the third person probably about 1800, fresh from another chronometer controversy – with Thomas Mudge this time – had this to say:

He always allowed Mr. Harrison's great merit, as a genius of the first rate, who had discovered, of himself, the causes of the irregularities of watches, and pointed out the means of correcting these errors in a great degree, in the execution of a portable time-keeper, of a moderate size, to be put on board of ship, not liable to disturbance from the motions of the ship, and exact enough to keep time within two minutes in six weeks. He made no opposit[ion] to Parliament granting him the remainder of the reward of £20 000; but only to the Board of Longitude doing it; as he had not submit[ted] to trials, and those sufficient to enable the Board to give it to him according to the terms of the Act.[28]

In July 1775, a year after Harrison received his final award, and some nine months before his death at the age of 83, Cook arrived home from his second circumnavigation, full of praise for Kendall's copy of the prizewinner which had been going impeccably for the whole of the three-year voyage, having maintained an almost constant gaining rate of between 9 and 13 seconds per day since arriving in New Zealand in April 1773. Writing to the Secretary of the Admiralty from the Cape of Good Hope on the last leg of his voyage, Cook said: 'I have received every assistance I could require from Mr. Wales the Astronomer; Mr. Kendals Watch has exceeded the expectations of its most Zealous advocate and by being now and then corrected by Lunar observations has been our faithfull guide through all the vicissitudes of climates.'[29]

The principal award offered by the Longitude Act of Queen Anne having been made to Harrison – nearly – Parliament thought fit to repeal all previous Longitude Acts and to pass another – 14 George III c.64, which became law on 2 June 1774 – halving the awards for all methods, and, for trials of chronometers, stipulating that two instruments of the same construction must be tried simultaneously for twelve months at the Royal Observatory, followed by two voyages around Great Britain in opposite directions, and such other voyages to different climates as the Board might decide upon.

In the event, it was not lunar distances but chronometers which eventually became the standard method of finding longitude at sea, but this did not become commonplace until the 1820s or later, by which time, thanks largely to the early work of John Arnold in mass production methods and quality control, they became cheap enough at least for foreign-going ships to have one or more.

The remainder of the story of this very busy decade as it affected the Commissioners of Longitude – and none more busy than Maskelyne – must be briefly told. Soon after the passing of the new Longitude Act, Thomas Mudge, inventor of the lever escapement (still used in mechanical watches today) and one of those who attended Harrison's discovery of H4 in 1765, presented his first chronometer and it was put on trial in the Great Room at Greenwich in December 1774. Although it promised well, it stopped after three months, and again a month later, this time with a broken mainspring – probably because it was carried up and down to the Transit Room every day for comparison with the transit clock. Mudge complained to the Board about this (and about the wording of the new Act) on 27 May 1775, so it was given a new trial starting in November 1776, this time in the Transit Room, which went on until 1778. Initially, Maskelyne spoke extremely highly of it. Meanwhile, Mudge had made two others, exactly alike, which he called 'Blue' and 'Green'. In 1779 and 1780, Maskelyne put these two on trial with pocket chronometer no. 36 of Arnold (called no. 39 in the Board minutes), and a chronometer by William Coombe.

Maskelyne was responsible for the scientific arrangements for two more voyages of exploration:

(a) in June 1773, the bombs *Racehorse* 8, Capt. the Hon. Constantine John Phipps (later Lord Mulgrave), and *Carcass* 8, Capt. Skeffington Lutwidge (with the 14-year-old Midshipman Horatio Nelson on board), sailed from the Nore 'towards the North Pole': the Board sent Israel Lyons (one of the almanac computers) in the *Racehorse* with Kendall's £200 chronometer K2 and a box chronometer by Arnold: in fact, Phipps's pocket chronometer by Arnold performed better than the Board's two chronometers;

(b) in July 1776, Cook sailed on his third voyage, this time to explore the north Pacific: the Board appointed his First Lieutenant, James King, their official observer in the *Resolution* with K1, and William Bayly in the *Discovery* with the £100 K3.

Two inventions which revolutionized the making of graduated instruments were submitted to the Board by Jesse Ramsden – his engine for

Fig. 11.2. The Maskelyne Crest and Arms. *Arms* – Sable, a fess engrailed or, between three escallops argent. *Crest* – a demi-lion sable, holding between the paws an escallop argent. *Motto* – Bis vivitur virtute. From the memorial tablet in Purton Church.

dividing circles and arcs of circles in mathematical instruments (particularly Hadley quadrants and sextants) in June 1774; and his engine for dividing straight lines in June 1777. Bird examined a sextant divided by Ramsden's engine and found that the graduations over the whole 120° were accurate to one two thousandth part of an inch, a remarkable result compared with previous methods used for making such divisions.[30] For instructing other workmen, writing illustrated descriptions (published by the Board in 1777 and 1779), and allowing the engines to become the property of the Board, Ramsden received £615 in 1775 for the circle engine, and £400 in 1778 for the straight-line engine.

In this decade also, there were important happenings between Maskelyne and the Royal Society. But these will be dealt with in the next chapter.

Account book entries

- 1773 Aug. 6 to 8 – Journey down to Ld. Clive's Oakly Park Shropshire; to Worcester coach £1 15s, thence in post chaises £1 5s 8d – £3 0s 8d.

  (A/C 1/36)

- 1776 Nov. 11 – Recd. of my Servant Wm. Crocket what he sold my Dunn horse for at Smithfield market – £5 10s 0d.

  (A/C 1/48)

- 1778 Feb. 15 – Pd. Mr Jno. Arnold for a seal of my arms cut & set in gold, with a motto Bis Vivitur virtute – £4 14s 6d.

  (A/C 1/52)

12

Weighing the World – Schiehallion, 1774

In his *Principia* of 1687, Isaac Newton pointed out that one of the consequences of his law of universal gravitation ought to be that, if a plumb-line were to be suspended close to a mountain, it would not be vertical because the plumb-bob would be attracted towards the bulk of the mountain. This deflection, said Newton, might be of the order of one or two arc-minutes – not very much, but quite enough to upset measurements taken with astronomical instruments which depended upon a plumb-line to give the vertical as a zero for measurements when very precise results were needed.

The French astronomers Bouguer and La Condamine were aware of this when they were measuring the length of a degree of latitude in Ecuador in 1738. They took good care to choose the site of their north-south line so that there were no mountains close enough either to the south, which might have deflected the plumb-line of the 2½ ft radius quadrant they were using to measure the latitude at each end of the line. They did, however, follow their measurement of a degree with an experiment to see whether the Attraction of Mountains (as this phenomenon came to be called) actually occurred. Setting up their quadrant close south of the great extinct volcano Chimborazo, they measured the apparent zenith distances of several stars. They then moved to the plain of Quito well to the west where Chimborazo could have no effect, and repeated the measurements with the same stars. There was definitely a deflection in the right direction, though the quadrant was not a precise enough instrument actually to measure that deflection. In his report, Bouguer expressed the hope that one day an experiment could be carried out under better conditions in Europe.

As we saw in Chapter 4, Maskelyne was well aware of this problem in St Helena in 1761, when he was using a 10 ft zenith sector (with a plumb-line) to try to measure the parallax of the bright star Sirius, which passes almost overhead there. St Helena is a mountainous island about 16 km from north to south and 13 km from east to west, rising to 800 m. To allow for the effect of the Attraction of Mountains on his measurements, he planned first to observe at Jamestown on the north coast of the island, where the plum-bob should be deflected to the south, and then to move his observatory to Sandy Bay on the south coast, where the plum-bobs should be deflected an equal amount to the north. In the event, the observations were frustrated by a design fault in the zenith sector, so no results were obtained.

The matter of the Attraction of Mountains was raised again in considering the measurements of a degree of latitude by Boscovich and Maire in the Papal States in 1750, Lacaille at the Cape in 1752, and Mason and Dixon in North America in 1766–8, all of which demanded very precise observation of the meridian zenith distances of stars to measure latitude on the ground. Commenting upon Mason and Dixon's report, Maskelyne said that there was 'no room for suspicion, that the plumb-line of the sector could be deflected materially from its proper position by the attraction of any mountain.'[1] Henry Cavendish disagreed, however. As a result of a mathematical investigation, he claimed that in Pennsylvania the deflection of the plumb-line could be affected not only by the attraction of the Allegheny mountains some 250 km to the north-west, but also by the *defect* of attraction by the depths of the Atlantic Ocean to the south-east. He said this could diminish the calculated length of the American degree by 60 to 100 Paris toises (a toise is the French fathom, 6 Paris feet, 1.949 m). He found similar effects for the South African and Italian measures.[2]

The Hon. Henry Cavendish, eldest son of Lord Charles Cavendish and grandson of the Duke of Devonshire, was a year older than Maskelyne and they had been contempories at Cambridge. Henry and his father, both distinguished experimentalists, were examples of a breed that was beginning to disappear, the natural philosopher interested in all things scientific, with ample private means. Henry joined his father on the Council of the Royal Society and as a member of the Mitre Club in 1760. One of the many scientific matters in which Henry interested himself was the matter of the mass and density of the Earth. He was intensely shy and reserved, apparently having no interests whatsoever outside matters scientific, almost never going to any function except Royal Society meetings and Joseph Banks's breakfasts. 'He probably uttered fewer words in the course of his life than any man who ever lived to fourscore years, not at all excepting the monks of La Trappe.'[3]

When the extra work imposed by the transit of Venus had died down, Maskelyne felt able to consider non-routine matters once more and turned his attention to the Attraction of Mountains, probably stimulated by Cavendish, whom he saw almost weekly at meetings or at the Mitre. (We have no evidence that they were not the best of friends, within the constraints imposed by Cavendish's shyness.) Perhaps a proper experiment could be done to settle the matter once and for all.

On 2 May 1771, Maskelyne informed the Royal Society's Council that he was going to propose this and asked that the Society's 10ft zenith sector, which he had found defective in St Helena, should be remodelled so as to be capable of measuring the meridian zenith distances of stars to the required precision of a few arc-seconds. The plumb-line suspension was to be made adjustable and a stand was to be provided to enable the instrument to be rotated into the plane of the meridian about a vertical axis, and to be turned through 180° to check the line of collimation (i.e. that the telescope's optical axis was pointing precisely vertical when the scale said it was). These modifications were taken in hand by the maker, Jeremiah Sisson. Unfortunately, the original instrument has not survived but the principle on which it worked is the same as that of Bradley's zenith sector, illustrated in Fig. 4.3.

In mid-1772, Maskelyne read a paper to the Royal Society entitled 'A Proposal for measuring the Attraction of some Hill in this Kingdom by Astronomical Observations'.[4] If the Bouguer experiment were to be repeated, he said, not only would it provide experimental proof of the theory of universal gravitation 'fit to convince those who will yield their assent to nothing but downright experiment', but it would also 'serve to give us a better idea of the total mass of the earth, and the proportional density of the matter near the surface compared with the mean density of the whole earth.'[5] (The internal constitution of the Earth – is it a hollow shell or is there a solid core? – was then a matter of scientific controversy.) Such an experiment, he added, would do honour to the nation in which it is made, and the Society which executes it.

The Council forthwith set up a Committee for Measuring the Attraction of Hills, comprising Maskelyne, Cavendish, Franklin, the Hon. Daines Barrington, and the Rev. Samuel Horsley. There was money left over from the £4000 granted by the King for the transit of Venus, so, with the King's approval, the Council decided to use this to fund the new project.

In the summer of 1773, Charles Mason was engaged – at half-a-guinea a day plus expenses – to do a tour of the Highlands of Scotland to search for suitable hills, and to look at hills in Yorkshire and Lancashire on return. Writing on 3 August to his friend Dr James Lind in Edinburgh,

Maskelyne tells him that Mason has set off on horseback with the necessary instruments – theodolite, chains, barometers, etc. – and will be calling upon him. 'We shall be much obliged to you for any information. . . Either an oblong hill (that has not been a volcano), or a long valley, the long way of either running from East to West or not deviating 30 or at most 40° from that direction, & ½ mile high or deep or nearly so, the hill with as few scarps & hollows as possible, would be proper for the purpose.'[6]

On 24 January 1774, the Committee (by now augmented by Dr William Watson senior and Matthew Raper, and by Joseph Banks the next month) took note of Mason's report:

Perthshire afforded us a remarkable hill, nearly in the centre of Scotland, of sufficient height, tolerably detached from other hills, and considerably larger from East to West than from North to South, called by the people of the low country *Maiden-pap*, but by the neighbouring inhabitants, *Schehallien*; which, I have since been informed, signifies in the Erse language, *Constant Storm*: a name well adapted to the appearance which it so frequently exhibits to those who live near it, by the clouds and mists which usually crown its summit. It had, moreover, the advantage by its steepness, of having but a small base from North to South. ...[7] (Fig. 12.1)

The obvious person to carry out the experiment was Mason himself, and the Committee offered him a guinea a day plus the costs of travel and carriage of instruments. Perhaps he felt the miserly terms were not good enough. Perhaps he was just tired of astronomy in the field, after the Cape of Good Hope and St Helena in 1761, North America in 1763–8, Ireland in 1769, and the Scottish Highlands in 1773. Whatever the reason, he refused – which, with hindsight, was sad from his point of view.

Maskelyne then suggested the name of Reuben Burrow, who had been his Assistant until the previous September. Burrow accepted. This was in January 1774. At their meeting in May, however, the Society's Committee of Attraction had second thoughts and begged Maskelyne to direct the operation personally, with Burrow as his assistant. Maskelyne reluctantly agreed, subject to permission from the King to absent himself from the Royal Observatory.

In Ecuador, Bouger had measured the deflection due to attraction by making two sets of zenith distance observations, the first close to the south side of the mountain, the second on the same parallel of latitude several miles to the west, well away from the mountain's influence: the

Fig. 12.1. Schiehallion, Perthshire, from the west. Schiehallion was the mountain chosen for the Attraction of Mountains experiment largely because of its saddle-back shape – its long axis runs some 4 km nearly east and west and, as can be seen from this photograph, its cross-section is regular in shape. Furthermore, there are no mountains or hills of any size nearby. Maskelyne's observatories were on the north and south slopes, about 400 m below the saddle back. (Photo: author.)

difference between the two results gave the amount of deflection due to the mountain.

Maskelyne proposed a somewhat different method of measuring the deflection – illustrated in Fig. 12.2 – by comparing the *apparent* difference of latitude (measured astronomically) between two stations on opposite sides of the mountain, with the *true* difference of latitude (obtained by triangulation). The deflection would be *half* the difference between the observations – 'for the plumb-line being attracted contrary ways at the two stations, the apparent zenith distances of stars will be affected contrary ways: those which were increased at one station being diminished at the other and consequently their difference will be affected by the sum of the two contrary attractions of the hill.'[8]

The three principal operations were therefore:

1. To find by celestial observation the apparent difference of latitude between the two stations O and P.

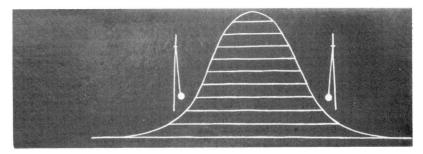

Fig. 12.2. The Attraction of Mountains. The plumb-bob of the zenith sector, which gives the zero of the instrument, is moved out of the vertical towards the mountain by the gravitational pull exerted by its mass. Observations of stars near the zenith on both sides of the ridge were used to show what that angle of attraction was, from which it is possible to determine Newton's universal gravitational constant, 'big G'.

2. To find by triangulation the true difference of latitude between O and P.
3. To determine the figure and dimensions of the hill, from which, knowing the deflection (the difference between 1 and 2), the density of the Earth can be determined.

Operation 1 was achieved by using the zenith sector to measure the zenith distances of a large number of stars within $8°.5$ of the zenith at both stations. Operations 2 and 3 were achieved by fairly conventional surveying methods, as can be seen in Fig. 12.3. Schiehallion rises about 2000 feet above the floor of the surrounding valley and of Lochs Rannoch and Tummel. Burrow sets up poles on little eminences around the foot of the mountain, forming a many-sided polygon, A, B, C, etc. Theodolite observations from these, from the observatories O and P, and from two cairns N and K built at each end of the ridge of the mountain, plus measured baselines both to north and south, gave the necessary information for Operations 2 and 3. The instruments used are listed in Appendix E.

Early in the summer, Burrow travelled to Schiehallion with the instruments, going by ship from the Thames to Perth up the River Tay, then about 45 miles by road to the mountain. By 30 June, when Maskelyne arrived, all was prepared: the triangulation stations had been selected and poles erected, and the southern observatory site was ready – the zenith sector and clock were erected in their 15 ft-square wood-and-canvas tent, the 1 ft astronomical quadrant was sheltered by a 5 ft diameter circular wall with a movable conical roof, the transit instrument was in a square

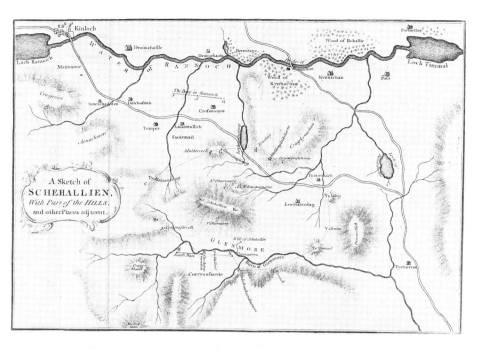

Fig. 12.3. Schiehallion and Surroundings. Maskelyne's observatories were at O and P, and the stations for the triangulation were placed on minor summits in the polygon $A, B, C \ldots U, X, S, R$. The Base in Rannoch, north of Crossmount is today under the waters of the reservoir Dunalastair Water. From *Phil. Trans. R. Soc.* 68 (1778), plate 11, p. 788. (Photo: RGO)

tent, and a temporary hut (called a 'bothie' in Scotland) erected for Maskelyne himself. Local help had been recruited, particularly William Menzies, a land surveyor.

Maskelyne himself was to make the astronomical observations, Burrow and Menzies most of the land surveying. Because of the weather, it was nearly three weeks before Maskelyne was able to set the sector precisely in the meridian and start proper observations. Then, from 20 July, he observed 34 different stars; on 1 August, he rotated the plane of the sector so that it faced west and re-observed the same stars – and more. About the 19th, the move to the northern observatory began, taking twelve men a week to move the equipment from O, some 1180 feet below the ridge on the south side, to P, 1450 feet below on the north side. Observations started again on 4 September, but because of appalling weather, they were not completed until 24 October. In all, Maskelyne made 337 observations of 43 different stars between July and October 1774.

Fig. 12.4. The Cairn Commemorating the Schiehallion Experiment, at Braes of Foss, an unmarked spot at right centre on Fig. 12.3, where the burn (or stream) crosses the road north of Leadnabroilag, not far from station Y (Velrig). It was set up by the Royal Society of London and the Royal Astronomical Society in 1982 to commemorate the 250th anniversary of Maskelyne's birth. (Photo: author)

Meanwhile, Burrow and Menzies were continuing the necessary land surveying. They measured baselines on flat ground both to the north and south, using them to triangulate the polygon surrounding the base of the mountain, which in turn was used to observe vertical sections from which its detailed shape and volume could be reconstructed.

Maskelyne returned to Greenwich at the end of October. Burrow remained until early frost and snow brought an end to operations in mid-November, leaving Menzies to complete in the summer of 1775.

Such are the bare bones of the story as given in Maskelyne's rather clinical account,[9] his only human touch being to list his various visitors on the mountain – James Stewart Mackenzie, Lord Privy Seal for Scotland (who lent a second theodolite and introduced him to his brother-in-law and local landowner, Sir Robert Menzies); Professors Alexander Wilson (and his son Patrick), Thomas Reid, and John Anderson of Glasgow; Mr Ramsay, Professor of Natural History at Edinburgh; Lord Polwarth; Messrs Patrick Copland and John Playfair from Aberdeen; Commissioner Menzies of the Customs in Edinburgh; Rev. Mr Brice; and Lt Col William Roy, Deputy Quartermaster General (with whom Maskelyne had travelled north in June), who had surveyed the Highlands of Scotland between 1747 and 1752 and was later to pioneer the Ordnance Survey.

What he did not mention was the drama of the fire in the bothie. There is a tale that has been handed down in Kinloch Rannoch and the surrounding Perthshire countryside that, when Maskelyne had completed his observations at the north observatory, he decided that he should give a farewell party on the mountain to thank the many local people who had helped him and Burrow over the last six months. He sent young Duncan Robertson, who cleaned and cooked for him in the hut, down to Kinloch Rannoch to collect the necessary supplies, including a keg of whisky. *Donnaeha Ruadh*, red-haired Duncan, was a fine fiddler, and his playing and singing had helped to pass many long evening hours during the summer and autumn.

It must have been quite a party, because the hut caught fire and Duncan's fiddle was destroyed. Duncan was heartbroken at the loss, but the story goes that Maskelyne consoled him, saying: 'Never mind, Duncan, when I get back to London, I will seek you out a fiddle and send it to you.'

And so it was. The new violin reached Duncan a few months later and he was so pleased with its tone that he composed a song, *A Bhan Lunnainneach Bhuidhe*, still included in Gaelic anthologies, of which these are a few verses in translation.

The yellow London lady

1. On the trip I took to Schiehallion
   I lost my wealth and my darling;

> There was not her like in Rannoch
> When she stirred herself to song
> *Though before I was mournful*
> *This year I am joyful*
> *I have received the yellow London Lady*
> *and she is the instrument for music.*
>
> 2. It was in Italy that she was reared;
> And she came across the sea to London,
> To give music to king and queen
> And to the high gentry of Europe.
>
> 3. She came of a precious family
> From the company of genteel royalty,
> From the high astronomer of the kingdom
> Who paid for her with his gold.
>
> . . . . .
>
> 12. It is Mr Maskelyne, the hero,
> Who did not leave me long a widower,
> He sent to me my choice treasure
> That will leave me thankful while I live.

The reference to Italy in verse 2 is explained by the fact that a paper label, now lost, with the words 'Antonio Stradivarius, 1729', was found in the original violin. Alas, it seems that the original fiddle was switched by an unscrupulous repairer in the nineteenth century, because the fiddle now in possession of Duncan Robertson's descendants appears to be of Edinburgh make of about 1840.[10]

The only indication of Maskelyne's own impressions found so far is contained in the rather pompous confidential letter already quoted, written sixteen years later concerning John Hellins, who was meanwhile 'holding the fort' at Greenwich:

> ... Mr. Hellins was my Assistant for 2½ years, between Michaelmas 1773 and Lady Day 1776, six months of which I was absent in the experiment which I made for the Royal Society on the attraction of the mountain Schehallien in Scotland. It was not expected that that experiment would have taken me above two months, but owing to the extraordinary bad weather it happened otherwise.
> My going to Scotland was a matter not of choice but necessity. The Royal Society, thinking that the person then actually employed by them in the surveying part of the business had been my Assistant could not be depended upon to

complete the work and publish the result, or at least, that the world would not be satisfied therewith on account of his inferiority of education and situation in life, made a point with me to go there to make the direction of the experiment, which I did, not without reluctance, not out of any wish to depart from my own observatory to live on a barren mountain, but purely to serve the Society and the public, for which I received no gratuity, and had only my expenses paid for me. The like objection to what was made to Mr. Burrough, as not being a person of weight enough to be trusted with the conducting of the experiment of Schiehallion, seems equally to weigh against the appointment of Mr. Hellins to the Professorship of Astronomy near Dublin.[11]

On 26 July 1775, Maskelyne presented to the Royal Society his paper giving the results of the experiment – final results as far as his own observations were concerned, preliminary results on the density of the Earth. His astronomical results gave the *apparent* latitude difference between the two observatories as 54.6 arc-seconds; land survey results said the true latitude difference was 42.94 arc-seconds. The difference between these two quantities represented the sum of the two contrary attractions of the hill. In short, Schiehallion deflected the plumb-line of the sector by 5.8 arc-seconds. He drew the following conclusions:

1. Schiehallion exerts an attraction that can be measured.
2. The force of gravitation varies (as Newton postulated) as the inverse square of the distance between the bodies, whether they be terrestrial or celestial [though he could hardly be said to have *proved* this].
3. The Earth's mean density is at least double that at its surface, implying that the density of the core is much greater than near the surface, 'totally contrary to the hypothesis of some naturalists, who suppose the earth to be only a great hollow shell of matter.'
4. The density of the superficial parts of the Earth is, however, sufficient to produce sensible deflections in plumb-lines, thereby causing apparent errors in the astronomical measurement of meridian arcs where a plumb-line is used to give the zero of measurement.[12]

The very next day, the Royal Society Council met with one member missing – Maskelyne. The proposal of the President, Sir John Pringle, that Maskelyne should be given the Society's highest award – the Copley

Medal for 1774 – was adopted *nem.con.* (The rule that the Copley Medal should not be awarded to a member of Council was on this occasion broken. He was dropped from the Council for a year in December 1776, but this may have had no connection.)

So, at the Anniversary Meeting on St Andrew's Day (30 November) 1774, Sir John presented the award:

These, Gentlemen, are the fruits of the operation of Mr. Maskelyne, during a residence of four months on a bleak mountain, and in a climate little favourable to celestial observations. To these inconveniences, however, he submitted with patience & complacency, as he went at your request and in pursuit of science . . .

Mean while we have the pleasure to find the doctrine of *universal gravitation* so firmly established by this finishing step of analysis, that the most scrupulous now can no longer hesitate to embrace a principle that gives life to Astronomy, by accounting for the various motions and appearances of the Hosts of Heaven.[13]

Although Maskelyne's report may have pleased the Royal Society, it did not please Reuben Burrow, who felt he had not been given the credit he deserved. Maskelyne 'ought to have mentioned that I had put all the Instruments and Drawn the Meridian line & put the instrument in order & did not know how to put it right again &c.'[14] As can be seen from his diaries, he was a jealous and resentful man with a chip on his shoulder, especially where rival mathematicians were concerned. Indeed, Augustus de Morgan had this to say about him: 'Reuben Burrow, an able mathematician but a most vulgar and scurrilous dog, left a diary, and notes in some of his books, containing much cursing, obscenity and slander.'[15]

Having had a dispute with the Royal Society in 1775 by refusing to hand over his original observation books and notes, stating that it was he and not Maskelyne that had done the major part of the work,[16] he published a long notice in the *St James's Chronicle* in January 1776, advertising the forthcoming publication of an account of the Schiehallion expedition, which was:

. . . designed for the Satisfaction of the Public in general, for the peculiar Advantage of those who desire to have the Reputation and Appearance of being learned, without possessing any of the necessary Qualifications.

At the same time will be published, Schehallien, a Poem, in the Stile of Ossian MacPhion; being a doleful Account of the perilous Perigrinations and dismal Disasters of the Sasnack Crean; intended as a proper supplement to the above.[17]

His description of the forthcoming works makes it obvious that they were to be a diatribe against the Royal Society and Board of Longitude in general, and Maskelyne in particular. Sadly for us, they seem never to have been published. It would have been interesting to see what was the merit of his case – and to read his scathing and perhaps slanderous comments.

The analysis of the results to ascertain the Mean Density of the Earth was entrusted by the Society to Charles Hutton, Professor of Mathematics at the Royal Military Academy, Woolwich since 1773. Hutton was five years younger than Maskelyne and the two men became great friends.

In his 1778 report, Hutton tells us that the methods that he used for this very complicated mathematical analysis were mostly due to Cavendish.[18] However, Hutton's use of land surface contour lines, universally employed in mapmaking today, seems to have been his invention, used here for the first time.

Having found the volume of the hill by these methods, he made the assumption that its average density was uniform and equal to that of common stone, 2.5 times that of water.[19] From this, he postulated that the density of the whole Earth was 4.5 that of water, significantly lower than Newton's guess of between 5 and 6.[20]

In 1798, Cavendish determined the Density of the Earth by an entirely different method, in the laboratory, using a torsion balance originally devised by John Michell, who had died in 1793. What came to be called the Cavendish apparatus consisted of two lead balls on either end of a suspended beam; these movable balls were attracted by a pair of stationary lead balls. Cavendish calculated the force of attraction between the balls from the observed period of oscillation of the balance and from this deduced the density of the Earth to be 5.45 times that of water. However, Hutton was convinced of the superiority of the mountain method and in 1811 persuaded John Playfair to do a thorough survey of the rock structure of Schiehallion, which allowed him to revise is own value up to 4.7. The currently accepted value is 5.52 – so Newton's guess could hardly have been closer.

Today, it is generally accepted that the mountain method has severe limitations due to the uncertainty of the density of the mountain and the smallness of the deflections which have to be measured. Nevertheless, at the time they were done, the experiment and its analysis were fine examples of all that is best in scientific research. One feels rather sorry for Charles Mason: if he had accepted the assignment when it was offered to him, he might today be known for the Mason experiment as well as for the Mason–Dixon line.

13

The 1780s – and a new planet

Personal and family

On Saturday 21 August 1784, at the parish church of St Andrew, Holborn, the Reverend Nevil Maskelyne, Doctor in Divinity, of Greenwich in the County of Kent bachelor, married Sophia Rose of the parish of Saint Andrew Holborn in the County of Middlesex spinster. They were married by the curate Mr G. Huddesford, in the presence of George Booth and Letitia Booth. The Bishop of London's special marriage licence of the previous day described Nevil as aged Forty Years and upwards (he was actually 52) and Sophia as Thirty Years and upwards (she was 32).

Meanwhile, the bridegroom had noted his expenses in his account book:

1784		£	S	D
Aug.20 Pd.	License for marrying Miss Sophia Rose	1.	11.	6
	Gold ring	0.	7.	0
	Rector's fee for marriage	0.	17.	6
	Made a present to the curate	0.	10.	6
	Clerk &c	0.	10.	0
		3.	16.	6[1]

A month earlier, on 20 July in the same church, the Reverend Sir George Booth, Baronet, of the parish of St John, Clerkenwell widower was married by Mr Huddersfield to Letitia Rose of the parish of Saint Andrew Holborn spinster, in the presence of Nevil Maskelyne and Sophia Rose.

Thus, it seems likely that Nevil met his future wife, co-heiress to a not inconsiderable estate, through Sir George Booth (1724–97), a distant cousin on his mother's side. Sir George has been presented to the living of Ashton-under-Lyme by his cousin Rt Hon. George Booth, Earl of

Warrington, Baron Delamere. He succeeded to the baronetcy on the death of another cousin Rt Hon. Nathaniel Booth, Lord Delamere, in 1770. Sir George was a magistrate and Deputy Lieutenant of the County of Middlesex. His first wife Hannah, daughter of Henry Taylor, had died in 1784 and he must have married Letitia very soon after.[2]

The two Rose girls, Letitia (1751-1823) and Sophia (1752-1821), were the only children and co-hieresses of John Pate Rose (born John Pate) (1723-58) of Cotterstock near Oundle in the County of Northampton and of the Island of Jamaica. Their mother was Martha Henn (1723-83) of a Staffordshire family, who eloped with John Rose. Though he never married Martha Henn, he accepted the two girls as legitimate and, when he died at the early age of 35 (the girls were only 7 and 6), he left very considerable property at Cotterstock and in London to Martha Henn, who henceforward called herself Martha Henn Rose.

When the girls were grown up, they and their mother seem to have divided their time between Cotterstock and a house in Milman Street, near Great Ormond Street, Holborn, not far from St Andrew's, which is presumably how they met Sir George (who lived in the neighbouring parish of Clerkenwell) and through him, Nevil Maskelyne.

Martha Henn Rose died on 22 October 1783, leaving Cotterstock to Letitia and dividing the rest of her estate equally between Letitia and Sophia.[3]

The Maskelyne's only child Margaret was born at Greenwich on 25 June 1785. She was christened the following day, being named after her god-mother and aunt, the Dowager Lady Clive. Nevil's first cousin Thomas Kelsall was god-father and Mrs Maskelyne the other god-mother.[4]

Meanwhile, Nevil's last surviving uncle, James Houblon Maskelyne, had died at Marlborough on November 8 1780, leaving a wife, Ann, and daughters Ann and Jane, aged 18 and 15 respectively, almost destitute. Assisted by his sister Margaret and cousin John Walsh, Nevil undertook to provide bounties for this family to the end of his life. In 1785 immediately after his marriage, he took Miss Ann Maskelyne to Calais to board with a Madame Le Maire, to be joined shortly after by her sister. They seem to have stayed in France until Christmas 1786. Jane married Joseph Toomer, ironmonger of Newbury in December 1787 (he was Mayor of Newbury in 1791-2), and they had eight children, the eldest daughter being named Margaret Maskelyne May, and the eldest son Samuel Nevil. Ann never married and both she and her mother received annuities from

Fig. 13.1. *Nevil Maskelyne*, 1785, aged 53, by Louis François Gérard Van de Puyl. From an oil painting signed and dated *L.F.G. Van de Puyl, 1785*, which hangs in the Royal Society's house in Carlton House Terrace, presented by Mrs Story (née Margaret Maskelyne) in 1830. Painted the year after his marriage, his account book for 29 November 1785 records that he paid Van de Puyl 20 guineas for the picture and £4 14s for a gilt frame. Margaret Clive said it was a very fine likeness (Lady Clive to Margaret Maskelyne, 30 October 1813). (Photo: Royal Society)

The 1780s – and a new planet

Fig. 13.2. *Sophia and Margaret Maskelyne*, 1786, by Louis François Gérard Van de Puyl. Oil painting in possession of Nigel Arnold-Forster, Esq. An entry in Nevil's account book for 24 May 1786 says: 'Mr. Vanderpuyl for Mrs. Maskelyne's portrait & the frame – 25.10.0.' Margaret Clive said this was a very fine likeness of Sophia. (Photo: Country Life Books)

Nevil's will in 1811: Jane had died in 1805 so Joseph received a bequest of £175 net. As we saw on p. 113, both Nevil and Margaret felt responsible also for William's putative daughter, now Mrs Jane Sacheverell. The sums involved in his philanthropy are summarized at the beginning of Nevil's second account book, covering the years 1785 to 1799, where he has made a short list called 'Outgoings' – Mrs Ann Maskelyne £20. 0. 0; Miss Ann Maskelyne £20. 0. 0; Mrs Sacheverell £30. 0. 0; Lying in Hospital £3. 3. 0; Marine Society £1. 1. 0; Philanthropic Society £1. 1. 0; Poor Orphans of Clergy £1. 1. 0; Society of Naval Architecture £2. 2. 0; Sunday School of Purton £1. 1. 0.[5]

Fellowship of Trinity College had to be relinquished on marriage. Possibly with this in view, Nevil was presented by his College to the living of North Runcton in Norfolk, being inducted in May 1782. At the same time, he resigned the living of Shrawardine. His tithes increased by nearly £100 a year, almost the same as his previous income from the Fellowship.

He was elected Fellow of the American Academy of Arts and Sciences at Cambridge, Massachusetts on 30 April 1788.

A seventh planet

On the evening of Tuesday 13 March 1781, William Herschel, organist and amateur astronomer, was working with a 7ft focus Newtonian reflecting telescope of his own design and making, in the garden of No. 19 King Street, Bath, continuing his systematic survey, started in August 1779, of all stars down to eighth magnitude (to isolate as many double stars as he could discover and use these data to measure stellar parallax). At 10.30 p.m., working with a magnification of 227, he saw what he described as 'a curious either Nebulous Star or perhaps a Comet.'[6] Changing eyepieces, he examined the object at magnifications of 460 and 932, recorded what he had seen in his observing book, and continued his search for double stars. The following Saturday, he saw the object again, realised that it had changed its place relative to the fixed stars, and concluded that it was a comet, not a nebulous star.

Herschel, born in Hanover in 1738, had first visited England in 1755 as an oboist in the Hanoverian Guards band. He decided that he should make a career in music in England, and in 1767 was appointed organist at the Octagon Chapel at Bath. From that date, he began to interest himself in astronomy, making his own telescopes, with stands of a novel design. From the outset, he found that just gazing at heavenly bodies did not satisfy him, and he became interested in what he called the Construction of the Heavens, the architecture of the system of stars of which the Sun is a member. It was to this end that his systematic survey of double stars was aimed: from it he hoped to be able to discover something about the distances of the stars from the Sun. He had already submitted papers to the

Bath Philosophical Society and was in touch with the Royal Society through Dr William Watson, MD, and his son Dr William Watson, Junior, both Fellows of the Royal Society.

On Sunday 18 March, five nights after he had first seen the object, he showed it to his friend Dr Watson, Junior, who suggested that a letter should be written immediately to his father in London, who would communicate the news to the Royal Society. This was done and, within a very few days, the news reached Maskelyne, who in turn passed it on to Charles Messier and Pierre Méchain in Paris. Herschel's letter was read at a Society meeting on 22 March.

Maskelyne first measured the position of the body on 1 April and it seems that it was he who was the first to propose that this was not a mere comet, but perhaps a planet. Six planets had been known since history began: could this be a seventh? On 4 April, he wrote to Dr Watson, Junior:

> I was enabled last night to discern a motion in one of them [of two bodies mentioned by Herschel], which as well as from its agreeing with the position pointed out by him convinces me it is a comet or new planet, but very different from any comet I ever read any description of or saw. . . This seems a comet of a new species, very like a fixt star. . . Astronomers are obliged to Mr. Herschell for this discovery, and I among the rest.[7]

Sir Joseph Banks had become President of the Royal Society in 1778, and was thus chairman of the observatory's Board of Visitors. The Astronomer Royal explained to his new boss on 16 April his frustrations over the failure of the equatorial sectors obtained in 1773 for just such an occasion as this:

> The Comet goes on very slow parallel to the equator according to the order of the signs at the rate of a little more than 2 minutes per day, so that its motion is accelerated very gradually. There never yet, I believe, was observed a Comet which moved any thing near so slow or for so long a time together. I found its apparent diameter last week 9 seconds. It had neither *coma* nor tail.
>
> I have unfortunately been obliged to make the best use I could of the equatoreal sector in the western building at the end of the terrace. The other, altered by Mr. Sisson, is not yet fit for use, but will be in a few days. Mr. Ramsden promises me a fine equatoreal instrument he has been so long about (5½ years) in a week's time; I wish he may keep his word. [He did not!]

Maskelyne then continues with his views on the discovery – views of a professional about an amateur – written just over a month after the discovery was made:

> Mr. Herschell is undoubtedly the most lucky of Astronomers in looking accidentally at the fixt stars with a 7 feet reflecting telescope magnifying 227 times to

discover a comet of only 3″ in diameter, which if he had magnified only 100 times he could not have known from a fixt star. I am afraid many such comets may pass by us without notice; for to count over 3000 stars (those in the British Catalogue) & examine whether no stranger has got among them, is not the work of a little time.

Perhaps accident may do more for us than design could; & this makes one wish that the number of astronomers was multiplied in order to increase our chance of new discoveries. Mr. Herschell has much merit in having ascertained the motion of this new star, and made good observations of its place.[8]

Herschel himself always maintained that the discovery was not an accident – '. . . as it was that day the turn of the stars in the neighbourhood to be examined,' he told Lalande, 'I could not very well overlook it.'[9]

Herschel visited London in May 1781, staying with Dr Watson, Senior, dining at the Mitre Club, and meeting Banks, Maskelyne, and Aubert, thus laying the foundation of firm and lasting friendships with all three.

In November, Herschel was presented by Banks with the Royal Society's Copley medal for 'the communication of his discovery of a new and singular star; a discovery which does him particular honour as in all probability this Star has been for many years, perhaps ages, within the bounds of Astronomick vision, and yet till now has eluded the most diligent researches of other observers.' He added that its amount of movement among the fixed stars since discovery was too small to be certain whether or not it was a primary planet, but it seemed probable 'that Mr Herschell has had the happiness of extending the bounds of the Solar System', because its orbit was apparently three times further away from the Sun than Saturn.[10]

In April 1782, King George III, prompted by Banks, expressed a wish to see Herschel. In May, therefore, Herschel packed up his 7 ft telescope and brought it to London, staying once more with Dr Watson in Lincoln's Inn Fields. At an audience, the King directed that in the first instance the telescope should be sent to Greenwich. Writing to his sister Caroline on June 3, William Herschel was cock-a-hoop:

The last two nights I have been star-gazing at Greenwich with Dr Maskelyne & Mr Aubert. We have compared our telescopes together and mine was found very superior to any of the Royal Observatory. Double stars which they could not see with their instruments I had the pleasure to shew them very plainly, and my mechanism [on the telescope stand] is so much approved of that Dr Maskelyne has already ordered a model to be taken from mine and a stand to be made by it to his reflector [the 6 ft Newtonian by Short]. He is, however, now so much out of love with his instrument that he begins to doubt whether it *deserves* a new stand.[11]

The rest of the Herschel story – his pension from the King, his abandonment of the musical profession and the move with his sister Caroline from Bath to Datchet and Slough (near Windsor), his many other researches and his large telescopes – has no direct connection with the Astronomer Royal or the Royal Observatory. Suffice it to say that the new star was soon definitely recognised as a planet. Writing to Herschel on August 8 1782, Maskelyne said: 'I hope you will do the astronomical world the favor to give a name to your new planet, which is entirely your own, & which we are so much obliged to you for the discovery of.'[12] The name that he did bestow upon it was *Georgium sidus*, a compliment to his new patron. Though never popular on the continent, the name 'the Georgian' continued to be used in Britain at least until 1847, when, in the Nautical Almanac for 1851, it changed to 'Uranus', a name used by J. E. Bode, the German astronomer, in the influential *Astronomisches Jahrbuch*.

The 1783 fireball

In the evening of 18 August 1783, a very bright meteor passed over Great Britain, the Low Countries and eastern France, travelling from north-west to south-east some 50 miles high. It was seen from Shetland, the Inner Hebrides, Edinburgh, Belfast, York, Windsor, Dunkirk (almost overhead), Brussels, Paris, and from near Dijon in Burgundy – about a thousand miles. (Bugge in Copenhagen told Maskelyne it was not seen in Iceland as the latter expected it might have been.[13]) Maskelyne (not yet married) was in Wiltshire at the time, but his domestic saw the meteor passing over Greenwich Park, rumbling like a cannon.[14] Among the many reports were those from astronomers Nathaniel Pigott near York and Alexander Aubert on horseback near Deptford. The artist Thomas Sandby happened to be on the terrace at Windsor Castle with Tiberius Cavallo, the physicist, Dr Lockman, and Dr James Lind MD, Maskelyne and Banks's friend from Edinburgh, now physician to the royal household, when they saw it pass from north-west to south-east. An aquatint showing the meteor, drawn by Thomas and engraved by his brother Paul, can be seen in Fig. 13.3.[15]

The exact nature of such bodies was still a matter for debate and this meteor, together with some others less spectacular in the next few months, caused great interest. On 18 September, Maskelyne reported the observations of Robinson of Perth and Pigott of York to the Mitre Club. The next month, Dr Charles Blagden, MD, no friend of Maskelyne's, wrote to Banks explaining that, many years before, Professor Winthrop,

Fig. 13.3. The Meteor of 18 August 1783 at Windsor Castle. This very bright fireball, seen in Great Britain, the Low Countries, and eastern France, was an object of particular interest to Maskelyne, who published a pamphlet on the subject on 6 November. By chance, the artist Thomas Sandby saw it on the terrace of Windsor Castle with several scientific friends. This aquatint by Paul Sandby after a drawing by his brother Thomas, was published in October 1783 and dedicated to Sir Joseph Banks. Without a title, it had the following legend:

The *Meteor* of Aug. 18. 1783, as it appeared from the NE. Corner of the Terrace at Windsor Castle, 18 Min. after 9 in the Evening: its apparent Diameter was nearly equal to the Semidiameter of the Moon, but its light much more vivid, its greatest Altitude was 25 Deg.

A [at left]   It's appearance soon after it emerged from a Cloud, in the NW. by W. where it was first discovered.
B [to the right of A]   It's further Progress when it grew more oblong.
C [above group on terrace]   When it divided and formed a long train of small luminous bodies, each having a tail: in this form it continued till it disappeared from the interposition of a Cloud in E by S.

(British Museum, Dept. of Prints and Drawings, case 261, 1872-7-13-481. Watercolour sketches in case 198, 1904-8-19-34 & 427.)

Maskelyne had two of these prints coloured, one for Lady Clive, one for himself. (NM to Dr. J. Lind (one of those on the terrace with Sandby), 11 Nov. 1783, R. Soc. MS 244/15.)   (Photo: BM)

of Cambridge, New England, had sent a paper to the Royal Society which was not printed, suggesting that meteors are terrestrial comets which excite light when they enter the Earth's atmosphere: 'Pringle took his ideas from it, which Maskelyne is now going to hash up warm.'[16] Blagden told Banks that, at the Club, 'Maskelyne was very peevish with Mr

Cavendish, almost rude, telling him it was evident he had not considered the subject'.[17]

It seems likely that Maskelyne did indeed wish to publish a definitive paper on the subject. In November, he sent out 100 copies of a three-page pamphlet 'A Plan for observing the Meteors called Fire-balls' (dated Greenwich, 6 November 1783, printed at his own expense, £3. 17. 6) which explained exactly what should be done by anyone seeing a fireball – who should then send their observations to him, the Astronomer Royal. To his chagrin, however, he was completely upstaged by a Blagden, who, at the Royal Society on 19 February 1784, read a paper called 'An Account of the late fiery Meteors; with Observations'.[18] Maskelyne never published on the subject.

The Paris–Greenwich connection

In October 1783, a month after the Treaty of Versailles had brought an end to the War of American Independence (and thus to war between Britain and France), the French Ambassador delivered a memoir written some years earlier by César-François Cassini de Thury, the noted cartographer and geodesist who was director of Paris Observatory. He claimed that there was still significant doubt as to the relative positions of the Royal Observatories at Paris and Greenwich – doubts of as much as 15 arc-seconds in latitude and 11 seconds of time in longitude – and suggested that these doubts could be resolved by an Anglo-French operation to connect the two observatories geodetically. The French had already observed a series of triangles between Paris and Boulogne, he said. what was now needed were observations to connect the French and English coasts – which would be a cooperative effort – and then a trigonometrical survey by British observers to connect Dover with London and Greenwich.

After consulting the King, the Secretary of State passed this memoir to the President of the Royal Society who in turn passed it to William Roy, by now a Major General, asking him to direct such an operation. Back in 1763, a decision had been taken that a general survey of the whole country should be carried out at public expense, but there has been no progress. Perhaps now was the moment to reconsider the project, to begin what today is known at the Ordnance Survey.

Early in 1784, Jesse Ramsden was commissioned to make a new theodolite with a horizontal circle 36 inches in diameter, which could be read to 1 arc-second, and an achromatic telescope of 2½ inches aperture. When eventually set up, it was estimated that the probable error for single observations was as little as 2 arc-seconds for distances up to 70 miles.[19]

Roy commenced operations in the summer of 1784 with the measurement of a base line on Hounslow Heath – near today's Heathrow airport – using standard glass rods made by Ramsden. In this he was assisted by Cavendish and Blagden, with occasional visits from Banks, encouraged by the King who visited the party on 21 August (Maskelyne's wedding day). It had been hoped that the main operation could be started in 1785 but, to Roy's frustration, Ramsden took more than three years to complete the theodolite.

In the meantime, Cassini de Thury had died, to be succeeded by his son, Jean-Dominique Cassini. (In 1768, he had sailed in the *Enjouée* frigate, 28, to try out the chronometers of Le Roy and Berthoud.) Enthusiastic personal contacts were established between Banks and Cassini in 1784.[20]

On 28 April 1785, a year after he had received it, Banks sent Cassini's memoir to Maskelyne to ask for comments so that they might be published. On the latitude discrepancy based on observations in the 1750s by Bradley and Lacaille, Maskelyne believed that Cassini had been misled because he did not have access to Bradley's observations which could have showed how the latter's value for Greenwich was obtained, particularly as the values used for refraction. Based on a Bradley manuscript kindly provided by Hornsby at Oxford, Maskelyne reckoned that Cassini's 15″ discrepancy should be reduced to 4″.5, which agreed with his own observations since 1765.

As far as the longitude difference was concerned, we now know (though nothing was published at the time) that Maskelyne took very positive action, sending his Assistant, Joseph Lindley, to Paris by coach and boat on a 'chronometer run', possibly the first recorded instance of the use of such a method of finding difference of longitude. Surprisingly – because the whole thing must have taken a great deal of organizing – the only details we have are in a small notebook with a marbled cover, preserved at the Royal Greenwich Observatory, on folio 6 of which are tables in Lindley's hand headed 'Determination of the Long. of Paris per Mr Arnold's Watches.'[21] No copies of letters to the lenders of the chronometers or to Paris Observatory, nor any details of funding, have come to light.

The instruments involved, all made by John Arnold, were these:

Chronometers
Box watch – had been to Paris 19 months before
The King's watch No. 87

The 1780s – and a new planet

No. 68 – belonging to Edmund Everard Esq., of Lynn
No. 89 – large watch belonging to Alexander Aubert Esq.
Small watches
No. 35 – J. Lindley's own
No. 108
No. 94
No. 118

All the timepieces were compared with the transit clock at Greenwich on 20 September 1785, at Paris on 27 September, and at Greenwich again on 3 October. The mean difference of longitude between the two observatories found from the four chronometers was 9m 20s.5, from the watches 9m 18s.0, weighted mean 9m 19s.8. In his critique of Cassini's paper, which discussed previous astronomical measurements but made no mention of the chronometer measurements, Maskelyne said he believed that 9m 20s would be 'within a very few seconds of the truth'.[22] Roy's value derived from geodetic measurements in 1787 was 9m 19s.4. The accepted value today is 9m 20s.91.

Ramsden eventually delivered the theodolite in time for the first observation to be taken near Hounslow Heath on 31 July 1787. Maskelyne describes the operations at Greenwich:

> Aug.14 [1787]. The Scaffolding over transit room for Gen. Roy was begun, & completed Aug. 18.
> Aug.25. Gen. Roy first came here and gave orders about preparing a flag-staff to put over the transit instrument [see Fig.7.1].
> Aug.27. He came here & put it up & dined here, & went to Norwood to observe it.
> Aug.28. White lights were fired by the flagstaff.
> Aug. 29. The new Theodolite was brought down by Mr. Dalby & Mr. Bryce, who dined here.
> Aug.30. Gen. Roy came down. The instrument was set up. The Genl., Mr. Cavendish, Dr. Blagden, Mr. Dalby, & Mr. Bryce dined here.
> The General staid here till Saturday evening [Sept.1] when he went to town & returned on Monday morning early, & took down the instrument & removed it away at two o'clock to Shooters Hill Tower.[23]

Having measured further angles with the theodolite at the various stations in Kent, Roy and his companions met their French counterparts in Dover on 23 September. After two days of amicable discussion settling the details of the cross-Channel observations, the French astronomers – Jean-Dominique Cassini, Pierre-François-André Méchain, and Adrien-Marie Legendre – returned to Calais, accompanied by Blagden who was to act as British liaison officer in France. The cross-channel observations were successfully completed on 17 October. Roy's observations in England were not completed until well into 1788, the results being published in 1790.[24]

It is of interest that Roy borrowed Harrison's prizewinning chronometer H4 for use with pole star observations, the only time it was ever used operationally. He reported that its rate in 1787 was 9 seconds per day gaining, in 1788, 3.5 to 4 seconds losing.

The observatory

The defective equatorial sectors provided by Jeremiah Sisson in 1773 have already been mentioned. Sisson died in 1783 and the maintenance of the instruments at Greenwich fell to Edward Troughton, with Peter Dollond providing the optical expertise. The only significant change in the observatory's equipment during this decade was the acquisition of one of Herschel's 7 ft Newtonian reflectors in 1788. Herschel was an official Visitor from 1784 to 1809 and all who saw the new telescope were full of praise. Nevertheless, only nine observations of occultations, etc., were recorded between 1799 and 1832, as against 541 for the 46 in and 87 for the 5 ft achromatic refractor.

George Gilpin, who in 1776 at the age of 21 had succeeded Hellins as Assistant, left in 1781, to be succeeded by Joseph Lindley who stayed just over five years. He left in September 1786, to become purser of the *Henry Dundas* East Indiaman. Then, for a few years, Maskelyne had great trouble finding suitable assistants, largely because of the ridiculously low salary which was all he could offer. From September 1786 to July 1789, he had no less than eight Assistants, all of whom stayed less than a year, some being dismissed after only a few weeks. Maskelyne was most concerned and in his memorandum book in 1787 is a list of some fifty 'Persons proper to inquire of about an Assistant'.[25] On the opposite page is the following memorandum:

Qualities to be required for an Assistant
  May 19, 1787

To understand Arithmetic, Geometry, Algebra, Plane & Spherical trigonometry, & Logarithms; to have a good eye & good ears, be well grown, & have a good constitution to enable him to apply several hours in the day to calculation, & to get up to the observations that happen at late hours in the night. To write a good hand, and be a ready & steady arithmetical compute. If he knew something of Astronomy & had a mechanical turn so much the better. To be sober & diligent, & able to bear confinement. Age from 20 to 40.[26]

One of the temporary Assistants whom Maskelyne would have liked to have kept was the 21-year-old John Brinkley, who came in June 1787 to get observatory experience and left after just over four months to take his degree at Cambridge where he graduated as Senior Wrangler and Smith's Prizeman. Four years later, partly through Maskelyne's recommendation, Brinkley was appointed Andrews Professor of Astronomy at Dublin. He was appointed Royal Astronomer of Ireland in 1792 and became Bishop of Cloyne in 1826.

At last, Maskelyne found a suitable candidate in John Crosley, who arrived in July 1789 and stayed just under three years.

With the arrival of their daughter in 1785 and the resultant increase in domestic staff, the Maskelynes found the accommodation in Flamsteed House inadequate. In 1750, Bradley had had an extension built onto the south side of the house, comprising a single room on the ground floor with a kitchen below, all surmounted by a single large chimney (Fig.13.4). The Duke of Richmond, Master General of Ordnance, visited the house in July 1789 and authorized the rebuilding of Bradley's extension to give two extra rooms each on the ground floor and basement. Maskelyne himself laid the foundation stone on 2 October 1789.[27]

The normal observation programme continued unabated in spite of the high turnover of Assistants. Caroline Herschel proved herself a great discoverer of comets, all of which were subsequently observed and their places measured by Maskelyne himself. She discovered her first on 1 August 1786; later discoveries were recorded in December 1788, April 1790, December 1791, November 1795 and August 1797. Writing to William Herschel soon after the discovery of Caroline's first comet, Maskelyne said: 'I hope we shall by our united endeavours get this branch of astronomical business from the French, by seeing comets sooner and observing them later.'[28]

Maskelyne maintained a busy correspondence with astronomers and mathematicians all over the world. A pocket book has survived giving the postage he had to pay on all foreign letters received and sent between 1776 and 1786.[29] (He was probably exempt from paying postage on inland letters by virtue of his office.) It shows a total of 138 IN letters

Fig. 13.4. *The East Front of Flamsteed House*, 1794, from a watercolour by Mr Sinley in the possession of Nigel Arnold-Forster, Esq. The dome on the right contained Sisson's equatorial sector, the far turret contained Maskelyne's camera obscura. The present courtyard was enclosed not long after this drawing was made. (Photo: NMM)

and 80 OUT letters in those ten years, of which 51 IN and 36 OUT letters were with correspondents in Paris (particularly with Edmé-Sebastien Jeaurat and Pierre Méchain, editors of the *Connaissance des Temps* at that period), apparently quite unperturbed by France's participation on the opposite side in the War of American Independence.

Of the many visitors to the observatory, Herschel and Roy have already been mentioned. Giuseppe Piazzi, who was about to set up a new observatory in Palermo, came to England in 1787, principally to place an order for a 5ft altitude and azimuth circle and a 5ft transit instrument with Ramsden, as advised by Maskelyne. He came to Greenwich with an introduction from Lalande and actually made observations there. Lalande himself paid several visits in 1788 when he was in London visiting instrument makers. On 4 December 1787, Maskelyne gave a dinner party at Greenwich for the French astronomers Cassini, Méchain, and Legendre, who were in England to consult with Roy about the trigonometrical operation, and to visit instrument makers and Herschel. Dining with them were Cavendish, Piazzi and Ramsden.[30]

In the 1780s, Maskelyne gave advice on the equipping of two new observatories – to Patrick Copland at Aberdeen, and to James Archibald Hamilton at Armagh.

The Nautical Almanac and the Board of Longitude

In May 1781, the second edition of the Requisite Tables was published, edited by Maskelyne and with the Explanation written by William Wales, by now Master of the Royal Mathematical School at Christ's Hospital.[31] (One of the unsuccessful candidates had been Reuben Burrow, to the latter's chagrin.) As we saw on p. 94, the almanac and Requisite Tables between them provided all the navigator needed to compute a longitude observation except for trigonometrical and logarithm tables. The new edition rectified this omission by including a table of natural sines and 5-figure logarithm tables or natural numbers and trigonometrical functions. It also included a table by Wales for finding latitude by two altitudes of the Sun. Ten-thousand copies were printed, sold initially at 12s each.

Throughout the decade, the Nautical Almanac was being published about five years ahead, the number printed being increased from 1000 to 1250 from the edition for 1788, published in 1783. Stocks of the almanac for 1787 ran out and it had to be reprinted, as did many subsequent editions. In 1785, the main print run was increased to 2000. The pay for computers was raised by £5 to £80 per almanac in 1781, and to £100 in 1789.

Other publications by the Board in this decade included the observations taken on Cook's first and third voyages, those of the second having been published in 1777. But the most important publication was *Mayer's Lunar Tables Improved by Mr Charles Mason*, published in 1787. Mason had first presented these tables to the Board in November 1777, asking for a £5000 award. He was actually awarded £750, having received £517.10 for earlier versions. In 1780, he submitted an improved set of lunar tables, which were used to calculate the Nautical Almanac for the next thirty years. For these improved tables he received an extra £50 and his widow another £50 some years later. In 1786, he decided to return to America with his second wife and eight children. Sadly, he fell ill on the voyage and died in Philadelphia on 25 October 1786 at the age of 58, being buried in Christ Church cemetry, where Benjamin Franklin, whom he had known, was buried a few years later. His widow subsequently received an extra £120 from the Board.

In June 1786, Maskelyne read a paper to the Royal Society in which he said that it was possible that the comet of 1532 and 1661 would return in 1788 or 1789, being best seen in the southern hemisphere.[32] A few months later, the government decided to fit out a fleet of convict ships to establish a colony in New South Wales under the governorship of Captain Arthur Phillip, RN. Maskelyne saw that here was an opportunity to get observations of the comet. William Dawes, a Lieutenant of the Royal Marines in the *Sirius* 22 was recommended as an accomplished observer and Maskelyne suggested the Board should lend Dawes an outfit of instruments similar to those taken on Cook's voyages. In the end, it was Phillip who signed for them, together with Kendall's chronometer K1, cleaned and adjusted after return from Cook's last voyage.

The First Fleet sailed in May 1787. After passing the Cape of Good Hope, Captain Phillip transferred to the *Supply* armed tender, taking Dawes and K1 with him.[33] He pushed on ahead of the main fleet and landed at Botany Bay on 18 January 1788. Not satisfied with its situation, he examined Port Jackson, a harbour mentioned by Cook, and founded the city of Sydney on 26 January 1788.

Dawes set up Australia's first observatory on Point Maskelyne in Sydney Cove (now Dawes Point at the south end of Sydney Harbour Bridge). He kept up a correspondence with Maskelyne during his time in New South Wales, reporting the observations he made.[34] No comet was seen. As a result of strained relations with Governor Phillip, he sailed for England with the instruments in the *Gorgon* 44 in 1791. He became Governor of Sierra Leone 1792-4, Master of the Royal Mathematical School at Christ's Hospital 1799-1800, and was in Sierra Leone again 1800-7. His son, William Rutter Dawes, became a distinguished astronomer in the nineteenth century.

Maskelyne's name came briefly into Australian history again a few years later. Rowdiness in Sydney in 1796 led Governor Hunter to create four divisions of the town, one of which was named Maskelyne – though why is obscure. The name did not persist for more than a year of so.[35]

Dissensions at the Royal Society

Joseph Banks, whose interests were principally botanical and antiquarian, President of the Royal Society since 1778, soon found several matters in the organization of the Society which displeased him, particularly the practice whereby the Secretaries nominated candidates for Fellowship without consulting President or Council. He was determined that candidates that did not have his approval should not be elected and he openly – perhaps unwisely – made his views known on the merits of candidates. He had very little in common with mathematicians and astronomers and was accused of packing the Council with his own candidates, not of those disciplines.

Things came to a head at the Council meeting of 20 November 1783. Dr Charles Hutton, Maskelyne's mathematician friend who had done the Schiehallion calculations, was someone whom Banks found particularly tiresome. A friend of Banks's predecessor, he had been appointed the Society's Foreign Secretary, responsible for foreign correspondence, in 1779. At a Council meeting when he was not present, Banks accused him of neglect of duty in not answering three letters from abroad. As a way of 'letting down the Doctor easily'[36], Banks proposed to the Council that a rule should be made that the Foreign Secretary should reside in London, whereas Hutton lived in Woolwich, about 8 miles away. Maskelyne and Paul Henry Maty, the latter one of the Secretaries whom Banks particularly disliked also, said that Hutton should be given the chance to defend himself. However, Banks's motion was carried and Hutton formally resigned, publicly, at the next full meeting of the Society on 27 November.

Then, at the Anniversary Meeting on 1 December, Fellows were astonished to find that Maskelyne's name has been omitted from the list of Council for 1783–4, with no astronomer or mathematician to replace him. At the next Society meeting, Edward Poore proposed, with Maty seconding, that Dr Hutton should be thanked for his services as Foreign Secretary. Banks stood firm. 'Only the President and Council,' he said, 'know whether the duties have been efficiently performed or not, and I, who do know, am of the opinion that they have not.'[37] To Banks's chagrin, the motion went in Hutton's favour with a narrow majority of 33 to 28, so the President had to deliver – reluctantly – the formal thanks of the Society.

Thereafter, the dissensions began to polarize themselves into Men of Science versus Macaronis,[38] the first group being those Fellows who treasured the traditions of the great Sir Isaac Newton and who looked askance at the successor who tried, as they put it, 'to amuse the Fellows with frogs, fleas and grasshoppers'.[39] Their self-appointed leader was Dr Samuel Horsley, a mathmatician and former Secretary of the Society who later became Bishop of St Asaph. Sadly for him, though much of what he had to say was absolutely right, his aspirations towards the Presidency – which it seems likely was his aim – were negated by his intemperate speeches and often unfounded charges. On 18 December, he moved that Hutton should be allowed to read a defence of his conduct. Seconded by Maskelyne, the motion was carried and Hutton's paper read, once again to Banks's chagrin.

But Banks, however much he might scorn national politicians of all colours, was no mean politician himself within the Society. At a general meeting of 12 February 1784, attended by 170 instead of the usual 70 Fellows (the result of a 'three-line whip' from the President), he received a majority of 119 to 42 in a vote of confidence proposed by Henry Cavendish, and motions 'to condemn the President's interference in elections' and 'to condemn him in influencing the votes of officers' were lost by 115 to 27, and 112 to 23 respectively, Maskelyne presumably voting on the losing side.[40]

At a later meeting, Maty resigned as Secretary – which is what Banks had been waiting for. Two candidates offered themselves: Banks's own choice Dr Charles Blagden, the Macaroni candidate, who had, so to speak, been waiting in the wings; and Hutton, the men-of-science candidate. Blagden was elected by 136 votes to 39.[41]

The effects of this rather unpleasant affair – of which it has been possible here to give only a bare outline[42] – lasted a long time, particularly between Banks on the one hand, and Hutton and Maskelyne on the other. (Horsley took no further part in Society affairs). Olinthus Gregory, Hutton's successor at Woolwich, gave several examples of Banks's spite towards the other two in a vicious polemic in the *Philosophical Magazine* after Banks's death – but he was very biased.[43] The fact remains that, as far as relations between Banks and Maskelyne were concerned, letters between them, couched in the third person during the 'affair', soon reverted to the first person, and Maskelyne came back on Council on 1 December 1784, being re-elected every subsequent year until his death in 1811. Both with strong personalities, contemporaries speak of continual friction between them at meetings of the Royal Society and Board of Longitude, but this is certainly not reflected in the minutes, where they seem more often to have been in agreement than otherwise. Whatever may have been their differences in committee, their personal relations seem to have been more cordial than might have been expected: in society, they appear to have enjoyed each other's company and met as friends and intimates.[44]

On 30 November 1780, the Society moved its premises from Crane Court, Fleet Street, to the newly-completed Somerset House in the Strand, some half a mile to the west. This was the first building in London designed specially to house goverment offices – as well as the Royal Society and the Royal Academy, on whose apartments the architect Sir William Chambers lavished particular care.

The Mitre Club moved from the Mitre tavern in Fleet Street to the Crown and Anchor over against St Clement's Church in the Strand (near Somerset House) at the same time. Its official title at that time seems to have been the Club of the Royal Philosophers. The name Royal Society Club began to be used in official records in 1794.

Soon after the move – and soon after the dissensions – we have the first reference to the Friday Club, a dining club for mathematicians which met on alternate Fridays. In October 1784, Blagden told Banks that Hodgson, a member of the Friday Club, had shot himself because of financial troubles, 'a fit end in my opinion of that sort of traitor to his country.'[45] (What Hodgson had done is not clear.) In December 1791, Maskelyne wrote to Herschel: 'I dined at our mathematical club last Friday, with Baron Maseres, & Dr. Hutton & Mr. Morgan . . . & Col. Etherington. . . We meet every fortnight regularly, dinner at half past four. . .'[46] In his next letter, he asked Herschel to dine with the club at the Globe Tavern, Fleet Street, on 13 January. The letter reproduced on p. 193 indicates that it was still flourishing in 1801.

Memorandum book extracts

[*1784 – No indication of source*]
American Academy of Arts & Sciences at Cambridge near Boston
Society of Arts & Sciences at Philadelphia.
College at Philadelphia. Dr. Ewing President.
University called Harvard College at Cambridge near Boston.
    Mr. Willard Prest. Secr. to American Acad. of Arts.
Yale University at Newhaven in Connecticut. Dr. Styles President.
An University in the Jerseys [later Princeton University].
An University in New Hampshire called Dartmouth College.
An University in the back parts of Pennsylvania about 120 miles from
    Philadelphia called Dickenson College.

(MB.2D/12^v)

[*1787*] Lake Champlain freezes sufficiently hard in winter to bear Carts and Artillery, is one of the best places that can be found any where for

measuring the length of a degree of Latitude. It is situated lengthwise nearly North and South, and lies between 45 and 43°. Mr Glenie.

(MB.2D/35)

May 1 1789. Recd. 8 doz. of Sherry of Dr. Shepherd. Vintage of 1782, the grapes grown truly and wine made within the district of Xeres, and such wine as cannot again be got, there being no more so old. Came to £9.10 in all, including bottles, deducting bottles at 2½d, so remains £8.10, which is at rate of £1.1.3 per dozen, or 1S 9¼d per bottle.

(MB.2F/17ᵛ)

14

The 1790s

Personal and family

At the beginning of the decade, Nevil Maskelyne had been Astronomer Royal for 25 years. He was 57, his wife Sophia 20 years younger, and his daughter Margaret 4½. His health seems to have been remarkably good although it is noticeable that his memorandum books from about 1790 first begin to include items concerned with health, not only about Margaret's childish ailments but also concerning those of Nevil himself – descriptions of symptoms and dates of occurrence, medicines prescribed and treatment prescribed, etc. In those same memorandum books and in the account books we see something of Margaret growing up – first the nursery maids, then the governesses, with teachers of reading, writing and arithmetic, then French, Italian, dancing, drawing – all in Greenwich, mostly at the observatory itself. By the time she was 16, she was an accomplished watercolourist (see Fig.14.1) and was regularly writing letters in French and Italian to Aunt Margaret, the Dowager Lady Clive.

Once a year in September or October, the whole family took itself for five weeks or so to Purton Stoke, to be looked after by the tenants, Mr and Mrs Large. While there, visits were paid to the tenants of the various other Maskelyne properties, Basset Down, Purton Down, Braden Farm, and to Lady Clive in Shropshire. Once a year also, Nevil would go to preach at North Runcton near King's Lynn, generally combining this with a visit to Sir George Booth and Letitia, Sophia's sister, at Cotterstock in Northamptonshire. (Sir George died in 1799 but Lady Booth continued to live at Cotterstock.)

Catharine, Edmund Maskelyne's widow, had re-married (for the second time) in 1785 but died with following year, whereupon the property of Basset Down reverted to Nevil Maskelyne, who eventually received the deeds from Catharine's relict, Mr Halkhead on 21 August 1792.[1] It remained let to tenants until his death in 1811, after which it became the family home for a hundred years or more.

Fig. 14.1. *The North Front of Flamsteed House*, 1801, from a watercolour inscribed *M. MASKELYNE. DELINT '01*, in the possession of Nigel Arnold-Forster, Esq. Margaret Maskelyne, an accomplished watercolourist, drew this at the age of 16. The steep path leading up to the Old Royal Observatory still keeps those who work there very fit. (Photo: NMM)

On 22 April 1799, the memorandum book records a visit by Mr Nevil Maskelyne, who had a brother Edmund.[2] It is possible, but by no means certain, that this Nevil was the grandfather of John Nevil Maskelyne (1839–1917), the illusionist, said to have been descended from Edmund, younger brother of the astronomer's great-great-grandfather, Nevil Maskelyne (1611–79).[3] In 1801, another visit is recorded: 'Sept. 24 Edmund Maskelyne Chilton near Swindon, called here with his brother Jasper who occupies the farm at Wootton Bassett.'[4]

Some time in the late 1770s, a drawing in chalk on blue paper was done of Nevil Maskelyne, seemingly his earliest surviving portrait. (Frontispiece) This has been attributed to John Russell, the portrait painter, who about that time became fascinated with the view of the moon through the telescope and turned to lunar mapping, measuring the positions of craters and other features with a micrometer, being advised by Maskelyne and Herschel. Several of his moon maps in pastel have survived.[5] (see Fig.14.3) In June 1797, he patented a moon globe on a special mounting, which he called the Selenographia.

Fig. 14.2. *Margaret Maskelyne*, by William Owen, RA, oil painting in possession of Nigel Arnold-Forster, Esq. (Photo: Country Life Books)

In 1805, in gratitude for the astronomical help he had received, he painted both Nevil and Sophia in pastel. (Figs.15.3 and 15.4) Shortly before his death in 1806, he was supervising the engraving of a pair of moon maps, which were eventually published by his son William in 1809, the first (with the rays of the Sun falling vertically as with Full Moon) dedicated to Maskelyne, the second (with the rays falling obliquely as

Fig. 14.3. *Joseph Banks*, c. 1788, by John Russell, RA. From a pastel in the Knatchbull Portrait Collection, first shown at the Royal Academy in 1788. Banks is holding a Moon map in pastel, several of which Russell drew with astronomical help from Maskelyne and Herschel. (Photo: Courtauld Institute of Art)

with successive phases) to Herschel. Although Russell gained much pleasure from his lunar mapping, he made no money out of it.

The writer of Maskelyne's biographical article in Rees' *Cyclopaedia* recounts this anecdote:

> He was fond of epigramatic thoughts and classical allusions; and even sometimes indulged in playful effusions of this kind, as appears by the following lines, which he composed on seeing Mr. Russell's selenographia, or map of the moon, executed with so much exactness.
> He makes Luna thus speak;
> > 'Me prope viderunt Actaeon, Endymionque;
> > Hos memini solos; ast ubi Russelius?'
>
> which he thus translated,
> > 'Actaeon and Endymion saw me near:
> > But when did I to Russell thus appear?'
>
> This epigram was composed extemporaneously when he was about seventy years of age, and is therefore the more worthy of being remembered as an instance of his lively and pleasant disposition at that advanced period. It also shews his passion for astronomy, which displayed itself so early in life, and which seemed to increase with his years.[6]

The observatory

Towards the end of the eighteenth century, astronomical instruments incorporating the full circle (as opposed to, say, the quadrant) became popular, particularly for theodolite-type instruments measuring both altitude and azimuth, such as Piazzi's 5-foot Palermo circle by Jesse Ramsden, completed in 1789. At the Visitation on 13 July 1792, Maskelyne suggested that a meridian circle of this type would be a valuable addition to the observatory equipment, whereby both the transits and zenith distances of bodies could be measured with the same instrument, and it would supersede the two mural quadrants (one 67, the other 42 years old) as well as the 42-year-old transit instrument. Though the Visitors approved, it was 1807 before Edward Troughton received an order for such an instrument.[7]

The only significant change in observatory equipment in this decade was that the 40-inch movable quadrant acquired by Bradley in 1750 (and almost never used) was, in 1797 with the King's permission, lent to Brinkley at Trinity College, Dublin. This he needed because Ramsden had failed to deliver the 8-foot meridian circle ordered as part of Dunsink Observatory's foundation equipment as long ago as 1785. (It was finally

Fig. 14.4. 8-Day Regulator Clock, by William Coombe, 1781, in a mahogany case. This clock was ordered by Maskelyne for the Grand Duke of Tuscany in 1781, for the latter's new observatory *La Specola* in Florence. According to Coombe, it gave him '... an opportunity of shewing my taste, etc., in the art...' (RGO MS 14/6). Unfortunately, after protracted tests at Greenwich, Maskelyne declared that its going was '... found not to answer, not being good enough to send...' (MB.2D), and in May 1788 he ordered another clock for the Grand Duke, from Larcum Kendall. Having already paid for the Coombe clock, however, Maskelyne decided to keep it for himself and had Kendall alter and set it up in Flamsteed House in October 1789. Thereafter Maskelyne kept a regular check on its going and his memorandum books show that it kept time to within a few seconds a week. Now at Belmont House, Faversham, the property of the Harris/Belmont Charity. (Photo: Harris/Belmont Charity)

delivered in 1807, finished by Matthew Berge after Ramsden's death in 1800. As an instrument maker, Ramsden was a genius, but he shared with other geniuses a disregard for timing. There is a story – surely apochryphal – that he arrived one day at Buckingham House precisely, he supposed, at the time named in the royal mandate. The King, however, remarked that he was punctual as to the hour and the day, but late by a whole year.[8])

An Assistant, John Crosley left in April 1792, to go to sea in the *Providence* on the Board of Longitude's behalf the following year. He was succeeded by Benedict Chapman, Joseph Garnett, and, on 23 May 1794, by David Kinnebrook, who was to achieve unwitting notoriety in the annals of experimental psychology many years later. In the *Greenwich Observations* for 1796, Maskelyne reported thus:

> I think it necessary to mention that my Assistant, Mr. David Kinnebrook, who had observed the transits of the stars and planets very well, in agreement with me, all the year 1794, and for the great part of the present year, began, from the beginning of August last, to set them down half a second of time later than he should do, according to my observations; and in January of the succeeding year, 1796, he increased his error of $8/10$ths of a second. As he had unfortunately continued a considerable time in this error before I noticed it, and did not seem to me likely ever to get over it, and return to a right method of observing, therefore, though with reluctance, as he was a diligent and useful assistant to me in other respects, I parted with him. . .
>
> The joint excellence of the clock, transit instrument, and method of observing, when properly attended to, have been the means of first discovering this error, and then ascertaining the quantity of it. . . I cannot persuade myself that my late Assistant continued in use of this excellent method of observing [introduced by Bradley], but rather suppose he fell into some irregular and confused method of his own. . . The great thing is to aim always at the truth, and avoid any partial method of observing . . . and then the independent observations of two observers will have a better chance of agreeing together, and with the truth. . . A good ear seems, in this kind of observation, to be almost as useful as a good eye.[9]

The incident was forgotten for many years (except perhaps by poor Kinnebrook himself) until about 1819 when Friedrich Bessel heard of it and began to make some investigations, suspecting that reaction time (or 'personal equation', as it came to be called by astronomers) could differ significantly from individual to individual. In the fullness of time, Kinebrook's misfortune led to what was virtually a new discipline, experimental psychology.[10]

The parting with Kinnebrook came on 12 February 1796 when

Thomas Simpson Evans, son of the mathematician Rev. Lewis Evans, became Assistant at the age of 19. He was the author of the account of an Assistant's life quoted in Chapter 7, and married Deborah Mascall, Margaret's governess, on 7 June 1798, leaving the Royal Observatory for William Larkins's private observatory in Blackheath at the end of the month. For almost a year, there was no permanent Assistant until the arrival of Thomas Ferminger, aged 24, in May 1799, who stayed until 1807.

One of Maskelyne's French correspondents was Vice Admiral le marquis de Chabert, a distinguished astronomer, cartographer and director of the Dépôt des Cartes et Plans of the French Admiralty. With the French Revolution, he came to England as an emigré. He was a frequent visitor to the Royal Observatory, and, as a Fellow of the Royal Society, was one of the official Visitors at the annual July Visitation every year from 1793 to 1801. Specifically, Maskelyne records a seven-week visit in 1797.[11] In his *Eloge* to Maskelyne, Delambre said how much Chabert appreciated Maskelyne's kindness to him during his enforced sojourn in London.[12] A biographical note of Chabert said this:

Becoming a refugee, he received in England the hospitality of the astronomer Maskelyne, who lavished all due care on an unhappy and distinguished colleague, and who offered to open to him an unlimited credit with his bankers, though the exiled scholar was not willing to avail himself of this.[13] [In fact, the name Chabert does not appear in any of Maskelyne's accounts with Messrs Thomas Coutts, his bankers.]

Chabert returned to Paris in 1802 and was appointed a member of the Bureau des Longitudes by Napoleon the following year. He died in 1805.

The Mudge affair, 1774–93

As we saw on p. 127, Thomas Mudge's first chronometer, made in 1774, was tried at Greenwich from November 1776 to February 1778. Maskelyne reported to the Board that it showed promise and Mudge was granted an advance of £500 to produce two identical chronometers, as the 1774 Act required. These were made in 1777, called 'Green' and 'Blue' after the colour of their cases. They were tried at Greenwich for 15 months in 1779–80 and again, after some modifications, for 14 months in 1783–4, but the Board were not satisfied they were exact enough to merit 'a great reward given by Act of Parliament'.[14]

In the mean time, Mudge's first chronometer had been acquired by

Count Brühl, ambassador in London of the Elector of Saxony since 1783, who entrusted it to Admiral John Campbell who was carrying out unofficial sea trials of chronometers by Arnold and others on voyages to and from Newfoundland, of which he was Governor. Both the Arnold and the Mudge performed excellently in voyages in 1784 and 1785.

Brühl then lent the Mudge to Franz Xavar Zach, at that time in London as tutor to Brühl's son. Zach, an arrogant young Austrian of boundless ambition, went so far as to claim in 1786 that the extreme accuracy of Mudge's watches had enabled him to discover serious inaccuracies in the Nautical Almanac. Bypassing Maskelyne, he sent these allegations direct to Hornsby and Banks. The former was so impressed that he obtained for Zach an honorary Doctor's degree: the latter arranged for the offending tables to be re-calculated by Michael Taylor, who completely vindicated Maskelyne and his computers. The same year, Hornsby proposed Zach as a Fellow of the Royal Society, but he was blackballed.[15]

Between May 1786 and July 1788, Zach took Mudge's chronometer travelling on the continent, including a sea voyage from Hyères to Genoa, when it was alleged to have performed impeccably. He then returned it to Brühl, who, with Zach, took up Mudge's cause with vigour in his disputes with the Board in general and Maskelyne in particular. Meanwhile, Zach had in 1786 been appointed by the Duke of Saxe Gotha to direct the new observatory at Seeberg, so, in 1788, he was once again proposed for the Royal Society, this time as a foreign associate: once again he failed to be elected.

Mudge submitted Blue and Green again in 1789 and they were tried at the Royal Observatory for the twelve months stipulated by the Act. On 4 December 1790, the Board received Maskelyne's report on the results of this new trial. 'It was thereupon resolved that as the said Timekeeper had not gone upon the Twelve Months Trial with the Exactness required by the Act of the 14th of the present King, the Board were not authorized to order farther Trial of them.'[16]

At their June meeting in 1791, the Board heard a memorial from Mudge, asking them to grant him whatever sum they thought he deserved for his invention and great labour. Whatever might be the technical merits of Mudge's chronometers, however, it was a fact that only one had ever been to sea and those tried ashore had, in the Board's opinion, failed to reach the required standard. On the other hand, John Arnold, for instance, already had more than a hundred of his chronometers at sea performing splendidly, and his and other makers' chronometers showed great promise in shore trials. Why should Mudge receive an award at this time? Why not Arnold?

By this time, Thomas Mudge's faculties were failing through old age but his lawyer son, also Thomas, took up the cudgels on his behalf, following almost the same tactics as Harrison had used successfully twenty years before.

Early in 1792, he published a 94-page 'pamphlet' entitled *A Narrative of the Facts, relating to some Time-keepers constructed by Mr. Thomas Mudge for the Discovery of the Longitude at Sea*... It also had a sub-title which is of particular interest here, . . . *together with Observations upon the Conduct of the Astronomer Royal respecting them*. In this vicious attack, Mudge accused Maskelyne:

(a) of persuading the Board and Parliament so to draft the Longitude Act of 1774 that it would be almost impossible for 'mechanics' to obtain any reward, because he, Maskelyne, had a personal pecuniary interest in such a reward going to himself as the architect of the tables for the rival lunar-distance method: 'He therefore of course cannot wish well to mechanics, who are candidates for the same prize, it being in his interest to have as few competitors as possible': Mudge claimed that Maskelyne had been heard to declare after the Act was passed, 'that he had given the Mecahnics a bone to pick which would crack their teeth';

(b) of receiving from the Board from time to time considerable sums of money for the construction of astronomical tables;

(c) of deliberately misusing Mudge's chronometers during trials;

(d) of using erroneous methods of calculation to prove how badly Mudge's chronometers performed; he was referring to Maskelyne's method of rating, which adopted the rate of going of the first month as the standard for the remainder of the trial;[17]

(e) of opposing Mudge on every occasion at Board meetings and influencing other Board members against him.

Although the main attack was directed against the Astronomer Royal personally, there were serious reflections on the Board as a whole. At their meeting of 3 March 1792, they passed a resolution deploring the attack on their colleague and specifically answering Accusation (b) above, affirming that all money paid to Maskelyne had been properly accounted for – in fact, it was almost all used for paying the computers of the Nautical Almanac – and that none had found its way into Maskelyne's own pocket.

Having thoroughly stirred things up and, with Brühl's help, done much lobbying in Parliament, Mudge the son presented a petition to the King, nominally in his senile father's name, praying for a reward for his invention of certain timekeepers, the one for which he had already received £500 and which had been to Newfoundland; and the two in which he was in dispute with the Astronomer Royal and the Board of Longitude on the method of calculating the performance. As the Harrisons had done, he finished off by an appeal to sentiment, praying that the House would 'grant such a reward for the Invention of them, and for the laborious Employment of the great Part of his Life, for the Benefit of the Public (the Petititoner being now in the seventy-seventh year of his age) as to the House may seem meet.'[18] On 21 February 1793, the petition was laid before the House of Commons by the Chancellor of the Exchequer, William Pitt, who was also Prime Minister.

Banks's indignation knew no bounds. Maskelyne had been criticised and this constituted a criticism of the Board itself, and thus of Banks himself. Any previous differences between them were completely forgotten.

The dispute Parliament was asked to decide was that between Mudge and the Board of Longitude, whose case is summarized by Banks in a letter to William Windham, leader of the pro-Mudge group who was to be Secretary-at-War from 1794 to 1801. (Having sailed with Phipps in the *Carcass* 'towards the North Pole' in 1773, Windham considered himself a navigational expert, though actually he had to be landed in Norway because of chronic seasickness.) After mentioning the 'unjust imputation of the Astronomer Royal having misrepresented his case to the Board of Longitude', Banks continued, 'I hold myself compelled by the duties of my station to give all the opposition in my power to the Bill, from the firm conviction that a Reward given to Mr Mudge by Parliament will operate as a reflection on the conduct of the Board and the character of the Astronomer Royal and also prove a discouragement to the diligence of better watchmakers who are now successfully employed in improving the construction of time-keepers.'[19]

Meanwhile, the petition was lying on the table of the House. On 24 April, Sir George Shuckburgh, amateur astronomer and friend of both Banks and Maskelyne, wrote to the former saying that Windham was proposing to name his own committee to consider the Petition, and that it would comprise Pitt for the Government, Charles James Fox for the Opposition, Gregor, Bragge, Rider, Windham, and Shuckburgh himself, the only pro-Board member.[20] Banks, who had no time for politicians, replied the following day that he had no fear of the committee 'except the ignorance which is very likely to manifest itself. I conclude Pit conceives a watch to be a thing composed of wheels within wheels like the Government of a Country, and thence deduces that he may throw new light upon it.'[21]

The petition was read a second time on 30 April, and was referred to Windham's committee which met in May, examining pro-Mudge witnesses – William Dutton (Mudge's partner), Aaron Graham (clerk to the late Admiral Campbell), and Count Brühl, who reported Zach's glowing report on the performance of Mudge's first chronometer. On 12 June, Windham presented to the House the Committee's report, which was ordered to lie on the table – where it remained for the remainder of the parliamentary session, thanks to energetic lobbying by Banks.

Meanwhile, the Board published Maskelyne's *An Answer to a Pamphlet. . .*,[22] a dignified if somewhat discursive (168 pages) reply to Mudge's *Narrative of the Facts. . .* In his attack on Maskelyne, Mudge had cited at some length Zach's evidence and opinions on the excellence of his father's chronometers. Replying, Maskelyne mentioned the unprovoked and unjustified attack Zach had made upon him in 1786 concerning the accuracy of the Nautical Almanac. He added that, while he had supported Zach's first attempt to be elected Fellow of the Royal Society in 1786, he had not felt inclined to do this again at the second – also unsuccessful – attempt in 1788.

Maskelyne concluded by asking a question. 'Shall this great and enlightened nation suffer an interested watch-maker, or the astronomer of a foreign prince, or any foreigner, to dictate to them the mode of trying time-keepers, when they have the Board of Longitude expressly appointed by Act of Parliament to consider these matters. . .?'[23] Not to be outdone, Mudge published on 3 July 1792 a 198-page *A Reply to the Answer . . .* in which he refuted all Maskelyne's statements, sentence by sentence.

There matters remained until 21 February 1793, when the petition was once more presented to the House by Pitt. It was once more referred to Windham's committee who examined the same witnesses as before, plus Tiberius Cavallo and Thomas Mudge Jr. Windham reported on 29 April, the House divided and, by 101 to 32, decided to refer the petition to a Select Committee – whose composition turned out to be nearly the same as Windham's two previous committees, but chaired by Sir Gilbert Elliott, the future Lord Minto.

The Select Committee first met on 9 May, appointing a technical panel to examine the chronometers – Samuel Horsley by now a Bishop, I. A. De Luc, and George Atwood, gentlemen; Jesse Ramsden and Edward

Troughton, instrument makers; and John Holmes, Charles Haley, and William Howells, watchmakers. They heard evidence from Graham and Mudge Jr on Mudge's behalf; from Banks, Maskelyne, Hutton, Wales, and three of Maskelyne's past assistants, Gilpin, Lindley and Crosley, on the Board's behalf. They also examined two chronometer makers, John Arnold and Josiah Emery, the former noting that he had made upwards of 900 timekeepers selling for 25 to 120 guineas, whereas Mudge had made only 4 or 5, vastly more expensive. Twenty of Arnold's were 'Number 1's' and he had accounts of the going of fifty.[24]

The Select Committee were not impressed with what the Board offered in evidence. They found in favour of Mudge and on 17 June 1793 the House recommended to the King that Mudge, upon making the Discovery of his Principles, should be awarded £2500, making £3000 in all.[25] Thomas Mudge, senior, died the following year.

The Nautical Almanac

At the beginning of the decade, the Nautical Almanac was being computed ten years ahead, using Meyer's tables as corrected by Mason for the Sun and Moon, and the tables in the second edition of Lalande's *Astronomie* for the planets. Then, in 1792, Lalande published a third edition containing much improved tables. Furthermore, the work of Pierre-Simon Laplace and Johann Tobias Bürg in France promised tables which were potentially capable of making the predictions in the Nautical Almanac more accurate still.

In 1793, therefore, at Hornsby's suggestion, Maskelyne proposed that calculations for the Nautical Almanac should be stopped for the time being – all almanacs up to 1804 were already done – so that advantage could be taken of the improved tables. The Board resolved to discontinue calculations for five years.

This imposed great hardship on the principal computers, three of whom – Henry Andrews, Mary Edwards, and Malachy Hitchins – presented petitions 'representing the great loss they should sustain if after having made the computations for 26 years they are deprived of an Employment from which they derive their support and on which they have been accustomed to depend, without having received any Notice, till lately, of this unfortunate event to them being likely to take place...'[26]

Over the next few years, Maskelyne took immense pains to do his best to alleviate their distress, finding work such as re-calculating some of the emphemeris from the new French tables. In the event, the main calculations were re-started after four years, in December 1797 for the almanac for 1805 (which was eventually published in 1801), using the astronomical tables for Sun, Moon and planets in the third edition of Lalande's *Astronomie*. For the 1806 and subsequent almanacs, Laplace/Bürg tables were used for the Moon, Lalande tables continuing to be used for the Sun and planets.[27]

In December 1799, the computers petitioned for a rise in salary because they said that Lalande's tables caused far more work than those previously used. The Board agreed to increase their salaries from £100 to £140 per almanac, starting with that for 1805, but they regretted they could not pay any compensation for the slack period since 1793.

The foundation of the French Bureau des Longitudes

On 7 Messidor, year III (25 June 1795), people's representative Grégoire (later to become Bishop of Blois) proposed to the National Convention in Paris that France should follow Britain's example by having a Board of Longitude. In a long speech he quoted Thermistocles as saying that whoever was master of the seas was also master of the world. The English had proved this to be so, he said, particularly in the war of 1761. And because of it, she had become a great power, whereas, by all the normal rules, she should play a mere secondary role in the political order. But now, said Grégoire, British tyranny must be stifled. And what better way than by using the methods she herself adopted? Britain had realised that, without astronomy, there could be no commerce, no navy. So she had gone to incredible expense in pushing astronomy to the point of perfection. And it was to her Board of Longitude that much of the credit for this was due: not only did it have enormous sums of money to disburse, but it also published the *Nautical Almanac* – admittedly on a French model – which had become their seaman's handbook.[28]

Grégoire suggested the foundation of a Bureau des Longitudes, on the British model but smaller and more manageable, to superintend the activities of Paris Observatory and the observatory of the Ecole Militaire, and to oversee the publication of the *Connaissance des Temps*, then, as now, under private proprietorship. So the Bureau was founded the same year, the founder members being the geometers Lagrange and Laplace; the astronomers Lalande, Cassini, Méchain, and Delambre; the retired navigators Borda and Bougainville; and the instrument maker Caroché, the professions of the members being laid down in the Law of 1795, which was passed immediately, thanks to the eloquence of Representative Grégoire.

The British Board was wound up in 1828 as it was considered that its original function, the discovery of the longitude, had been achieved. The French Bureau still survives and remains a most prestigious scientific body, acting as the French Government's chief advisers on scientific matters concerned with navigation.

Other Board of Longitude matters

The Board first had a Secretary in 1763, when an Admiralty clerk, John Ibbetson, was appointed to the Board part-time. He remained Secretary nearly twenty years, despite complaints by the Speaker of the House of Commons and others in 1779, when Maskelyne wrote to Banks complaining of Ibbetson's intolerable neglect of duty; he seemed 'to take a pleasure in plaguing every body who has any business with the board, & I myself am a principal sufferer from the connexions I have with the business of the board in appointing the computers of the nautical almanac, paying them & publishing the almanac & other works of the board.'[29]

Nevertheless, Ibbetson managed to retain his post until appointed one of the Deputy Secretaries of the Admiralty in 1782, when he was succeeded at the Board by Sir Harry Parker, Bart., also an Admiralty clerk. The latter resigned in March 1795, being succeeded in December by William Wales who was then at Christ's Hospital. Wales died in December 1799 and was succeeded in turn by Maskelyne's former Assistant, George Gilpin, who was by now Assistant Secretary to the Royal Society, and had had charge of the warehouse in which the instruments of both the Board and the Society had been stored since 1791. Gilpin died in 1810 and was succeeded by Captain Thomas Hurd, RN, Hydrographer of the Navy.

On the Board itself, other important changes occurred, the first since Banks succeeded Pringle in 1778. Antony Shepherd, Plumian Professor of Astronomy and Experimental Philosophy at Cambridge, died on June 15 1796, and the 44-year-old Samuel Vince was elected in his place. A great admirer of Maskelyne, he dedicated his *A Complete System of Astronomy* to him in 1797, 'from motives of private friendship' as well as a desire to pay tribute to his astronomical work. It was to Vince that Maskelyne bequeathed his scientific papers; his contemporaries hoped that some publication might result, but this did not happen.[30]

In December 1796, Abram Robertson was elected Savilian Professor of

Geometry at Oxford, in place of Dr John Smith, MD, who had succeeded Betts in 1766. In June 1794, Robertson had spent a fortnight at the Royal Observatory, gaining practical observing experience.[31] At Cambridge, William Lax replaced John Smith DD as Lowndes Professor in 1795, and Isaac Milner replaced Edward Waring as Lucasian Professor in 1798.

It was in this decade also that there occurred two examples of the great kindness which Maskelyne displayed toward those for whom he felt responsible. Michael Taylor, a mathematician about whom not a great deal is known, had been employed by the Board to calculate *A Sexagesimal Table* which was published in 1780. He had then, at Maskelyne's suggestion, started to prepare the Table of Logarithms to end all such tables, with logarithms to seven figures for numbers and five figures for trigonometrical functions, for which subscriptions at 3 guineas a book were sought in 1784, Maskelyne helping with the advertisement and collection of these.

When Taylor died in 1789 or 1790, the main work was finished but Maskelyne undertook to write the Preface and Precepts and to see the book through the Press. It was published early in 1793. But Maskelyne's help did not finish there. At the request of Thomas Taylor, Michael's father, Maskelyne and Francis Wingrave, the publisher and bookseller, became Trustees of Michael Taylor's estate, seeing to the schooling of young John Michael Taylor, who seems to have been motherless.[32]

The second example concerned William Gooch, a graduate of Gonville and Caius College, Cambridge, whom Vince had recommended should go to the Pacific with Captain George Vancouver. An expedition to the north-west American coast had been planned under the command of Captain Henry Roberts in 1790, but this was delayed by the Nootka incident with Spain and, when the expedition finally sailed in 1791, it was under the command of Vancouver, who had sailed with Cook on his second and third voyages. His job was to survey the American Pacific coast and do his best to ensure that no other nation, particularly Spain or the newly-independent United States, took possession of what is now Canada's Pacific coast.

Initially, Vancouver had said he did not want an astronomer but changed his mind shortly before he sailed in the *Discovery*, in company with the *Chatham* brig, in April 1791. That same month, Vince and Brinkley (who left the same week for Dublin) took Gooch to Greenwich to be vetted by Maskelyne. 'I find Dr M a very pleasant man,' he wrote to his parents, 'he doesn't seem to intend to examine me, but to be perfectly satisfied with Vince's word.'[33] Gooch stayed at Greenwich for a month or more, taking part in the work of the observatory, attending three dinners at the Royal Society Club as Maskelyne's guest.

At their meeting on 11 July, the Board appointed Gooch at a salary of £400 a year, lending him an outfit of navigational and astronomical instruments as well as a clock by Earnshaw, two Arnold chronometers, and an Earnshaw pocket watch. (Kendall's K3 had gone with Vancouver in the *Discovery*, Arnold's No. 82 in the *Chatham* with Broughton.) A few days later, Gooch joined the *Daedelus* storeship, commanded by Lieutenant Richard Hengest, which sailed independently late in August 1791 to meet Vancouver at Nootka Sound in today's Vancouver Island.

Some two years later, on 12 June 1793, John Pitts, who had sailed in the *Daedelus* as Gooch's servant, called on Maskelyne with some very grave news: on 10 May the previous year, before reaching Nootka, Hengest and Gooch had gone ashore with a watering party on Oahu, one of the Sandwich Islands (today's Hawaiian Islands), and had both been murdered by natives. The instruments and chronometers were safe and had been handed over to Vancouver.

Maskelyne immediately wrote a letter of condolence to Gooch's father in Norfolk, enclosing £215 as an advance on eleven months' salary due up to the day of the murder. According to custom, Gooch's effects had been auctioned onboard the *Daedelus* and Vancouver sent the proceeds, £178; but on applying for this from James Sykes, the ship's agent in London, Maskelyne was told that Vancouver's purser claimed 5% and he, Sykes, 2½%. Maskelyne felt that there was no legal justification for this: 'But justice should be done,' he wrote to William Gooch, senior, 'I have always heard that these ships agents are very sharp people.'[34] Over the next few months, Maskelyne took immense trouble in settling Gooch's affairs.

Meanwhile, the Board had decided to send a replacement for Gooch and, on 7 December 1793, appointed John Crosley, who had been Maskelyne's Assistant when Gooch was chosen instead of Crosley himself.[35] He was to sail in the *Providence*, which was going to join Vancouver with Captain William Broughton, who had commanded the *Chatham* and was then sent home by Vancouver in 1793 with dispatches. After a year's delay, the *Providence* sailed in February 1795, with five of the Board's chronometers, three by Earnshaw and two by Arnold. After a leisurely passage, they arrived in Nootka Sound to discover that Vancouver had already sailed for home, having put Maskelyne on the map again by naming an island and a point at the entrance to Portland Sound (just north of the

Queen Charlotte Islands) after him.[36] (On today's British Admiralty Chart 1637 of the New Hebrides will be found the Maskelyne Islands and Mount Maskelyne, so named by Wales during Captain Cook's second voyage.)

Broughton and his officers decided that the most fruitful thing they could do would be to survey the coast of China, to fill in the great gap in the world's charts between 30° and 52° North. Alas, disaster struck again, though Crosley was much luckier than Gooch. In May 1797, the *Providence* struck a reef east of Formosa (Taiwan) and was a total loss. Luckily, she was in company with a tender which Broughton had purchased and no one was lost. Crosley came home by East Indiaman.

The publication of Bradley's observations

As we saw on p. 122, a law suit by the Crown to recover Bradley's Observations was about to be started when his son-in-law presented them to Lord North who in turn presented them to the University of Oxford on the condition of their printing and publishing them. As a result, the Crown in 1776 abandoned the law suit though still claiming the right in law to possession of the papers. The University accepted this obligation and handed the papers over to Thomas Hornsby, Savilian Professor of Astronomy, for editing.

In November 1784 when there had been no news of progress for eight years, Banks, at Maskelyne's suggestion, mentioned the matter in passing in a letter to Hornsby. Hornsby said he was sorry but he had been delayed by ill health. Over the next few years, Banks, as President of the Royal Society and Visitor to the observatory, wrote several letters to Oxford suggesting that, as the Press could not possibly make any money out of publishing them would it not be better to hand them over to the Royal Society who could get them published at public expense? He asked also that pages might be sent to Maskelyne one by one as they were printed off.[37] There was no response.

Another six years passed. In 1791, Maskelyne, strongly supported by Banks, persuaded both the Royal Society and Board of Longitude that action must be taken. Banks wrote once more to the Vice Chancellor, to which he received a reply that the Delegates of the Oxford Press 'were not conscious of any unnecessary delay respecting the publication of Dr Bradley's papers'. It was just that Hornsby's health had delayed matters.[38] At the observatory Visitation of 29 July, resolutions were passed, copies of which were sent to Oxford, suggesting that Hornsby's ill health was

The 1790s

not a sufficient justification for a delay of fifteen years, but that someone else should have been engaged. For want of access to the observations, the public suffered considerably: science, the Royal Navy and the whole shipping interest of Great Britain, the Board of Longitude, the Royal Observatory – all of these were the losers.

On 22 December, Maskelyne himself wrote to the Vice Chancellor, asking that at least the observations could be sent to him sheet by sheet as they were printed. He said that he had access to all the original observations taken at Greenwich, excepting only Bradley's; 'the want of these makes a chasm of twenty years in the series from 1676.' He added that all his own observations up to 1790 had already been published and presented to the University: could not Oxford send Bradley's in exchange? He was told that a copy of the observations would be sent when they were published – whenever that might be.

At all the Board meetings where the matter was discussed, Hornsby was present as an *ex officio* member – but apparently did not speak. On 2 March 1793, however, the Board formally confronted him: in reply, he said that the first volume would be published on or before that day in a year's time. On 1 March 1794, he regretted his health had prevented his previous promise being achieved, but certainly the first volume would be printed by the December meeting, or he would give the task to someone else. On 6 December, he said he was sorry, but he had been ill.

A letter from Earl Spencer, First Lord of the Admiralty, to the Duke of Portland, Chancellor of the University of Oxford, produced almost no result, except that the University once more refused point blank to appoint anyone other than Hornsby to complete the task. In his covering letter enclosing the reply from the Delegates of the Press, the Duke said, rather ruefully, that his Oxford friends were of the opinion 'that the completion of the work is proceeding with all reasonable – I had nearly said all possible – expedition.'[39]

Maskelyne, Banks, the Society, the Board – all were in the last stages of frustration. On 6 June 1795, the Board ordered that copies of the Board minutes and related documents should be printed and distributed widely in the Government, in Parliament, and in the universities.[40]

At last, in March 1789 – just 36 years after Bradley's death and 22 years after Oxford had undertaken to publish – Maskelyne heard from Robertson, Hornsby's colleague, that the first volume of Bradley Observations was finished. 'I congratulate you,' he wrote to Banks, 'on this pleasing intelligence . . . I shall ever acknowledge your kind & public spirited assistance in this business, worthy of the President of the Royal

Society, and his accustomed zeal to promote every branch of Science.'[41]

On 3 May, Maskelyne received his copy of Volume I from Oxford,[42] giving transit observations from 1750 to 1755, quadrant observations from 1750 to 1758, and lunar places derived from Bradley's observations from 1750 to 1760.

In 1801, Abram Robertson, Savilian Professor of Geometry, was asked to edit the second volume covering 1756 to 1760. Published in 1805, that volume included, at the Board's request, the observations taken by Bliss and Green between 1762 and 1765, to provide a link between the observations of Bradley and those of Maskelyne, whose own first volume had been published twenty-nine years before.

Memorandum book extracts

Monday May 5. 1794. New Regulation of the penny post took place. It comes into Greenwich at 8 a.m. & 2 and 6½ p.m. & goes out a little before 10 a.m. and at 4 p.m.

May 29. It goes out at 9 & a little before 4.

(MB.2G/21^v)

April 24. 1794. Meeting of the County of Wilts to augment the militia by 400 men.

May 5. I ordered Mess. Coutts to subscribe 20 G for me at Mess. Hoares.

(MB.2G/22)

[September 1794, just before the annual visit to Purton Stoke]
Memoranda at going to the country.

Kearsley's abstract Tables of taxes;
Tea; Wine; brandy; Rum;
Buchan's domestic medicine;
Prayer books;
Wood's essence for meat & fish;
Flannel waistcoat;
spare wig;
Boots, spurs & whip;
oyl-skin hood;
Umbrellas;

knife with instruments;

Paterson's roads & maps of Counties;
Cork-screw; Decanters;
Lewis's Dispensatory;
Map of Wiltshire;
Worstead boot-stockings;
Spare suit of cloths;
great Coat;
gloves;
spare pair of shoes;
Paper, pen and ink, wax and wafers, & pen-knife;
money-weighing machine.

The 1790s

Rasor, strap, brush, shaving cloth;
Papers about the estates & leases of the estates;
maps of the seat of war;
2d pair of buckles;
wax-candle;
Cheshire & parmesan cheese;
ounce measure;
quills, pens;
silver tea spoons & table spoons;
silver scales;
Sermons, sermon-book & band;
Pocket-farrier;

Books of amusement;

Papers about the board of longitude
Messuage cards;
6S stamp & other stamps;
clothes brush;
funnel;
lamp wicks;
ivory folder;
pencils;

Chariot stool;
small quad. & compass and measuring wheel.

(MB.2G/26)

15

The final years, 1800–11

In the last few chapters, we have told the story subject by subject, not necessarily in strict chronological order. For this last chapter of Nevil Maskelyne's life story, we will, where possible, revert to chronological order.

A new London Bridge

In April and May 1801, Maskelyne was one of six mathematicians, five engineers, and two ironmasters – 'some of the persons most eminent in Great Britain for their Theoretical as well as Practical Knowledge of such Subjects' – to give evidence to the Parliamentary Select Committee upon Further Measures for the Improvement of the Port of London, to consider the revolutionary design for a new London Bridge by Thomas Telford – to be of cast iron with a single arch of 600 feet span, and 65 feet high. Maskelyne gave his opinion that the plan was practicable and advisable 'and when filled up with more minute particulars, and perhaps some further improvements, is capable of being rendered a durable edifice.'[1]

In the event, the design was abandoned because of its unprecedented scale and lack of technical knowledge, planning problems in the approaches, and the fact that docks being built down-river made such development less necessary. A masonry bridge with five spans, designed by John Rennie the elder, was begun in 1824 and completed in 1831.

The minor planets

For astronomy, the nineteenth century opened most auspiciously with the discovery of another planet on 1 January 1801, the first day of the new century – though the news took a month or so to reach a world then at war. The discovery was made at the observatory at Palermo in Sicily by Giuseppe Piazzi, and the story was succinctly told by Maskelyne after the planet had been rediscovered the following year:

A short account of the new planet, discovered by Mr Piazzi, Astronomer to the King of Naples and Sicily, and called by him *Ceres Ferdinandea*, in honour of his King and Patron whose name is Ferdinand, and in allusion to Ceres the ancient tutelary Goddess of Sicily, who first taught the use of corn there.

While Mr. Piazzi was making his observations for a new catalogue of fixt stars, with a 5 feet circle made here by the late Mr. Ramsden, he observed this star on 1 Jan. 1801 without knowing or suspecting it to be a planet; but by observations in the following days, he found it had a motion; he continued his observations of it for 6 weeks, when he fell ill, and the planet was lost to Astronomers. It appeared to him in Jan. 1801 to be of 8th Magnitude.

Mr. Gauss, an Astronomer at Brunswick, in Germany, calculated its orbit, by means of Mr. Piazzi's observation by the help of which Dr. Zach discovered it on 7th December last, at Saxe Gotha, and Dr. Olbers at Bremen on the 1st of Jan. of this year. It has been since observed at Paris on 24 Jan. and here on the 3rd and 4th of February.

It appears here as a star of 9.M[agnitude] no way different from a fixt star of the same magnitude. It will be in opposition to the sun on the 13th of March 1802. Its motion is between Mars and Jupiter. Its mean distance from the sun is 2¾ that of the mean distance of the earth. Its periodic time 4 years & 7 months. Its eccentricity a little less than that of Mars. Its inclination to the E[c]liptic $10°37'$.

Feb.17:1802

N.M.[2]

It was not until 23 January 1801 that Piazzi took any steps to communicate his discovery, when he sent letters to Barnaba Oriani in Milan, Johann Ellert Bode in Berlin, and Franz Xaver Zach (by now Baron von Zach) at Seeberg near Gotha. Because of war conditions, these letters took more than three months to reach Germany; but when they did arrive, they caused great astronomical excitement. According to the *Titius–Bode Law* (see Glossary), whose validity seemed to have been confirmed by the discovery of Uranus, there should be a planet in orbit in the apparently empty space between Mars and Jupiter and, in the 1780s, Zach organized an association of twenty-four astronomers (of which Piazzi was not one) systematically to search for it. Could Piazzi's discovery, which by now was too close to the Sun to be observable, have been the missing planet?

Exactly when or how Maskelyne heard the news is not clear, but the letter written by him reproduced below seems to imply it was in the summer of 1801. From the tone of the letter – for whom it was intended, we do not know – he was obviously angry not to have received the news direct from Piazzi, whom he had assisted and entertained at Greenwich in 1787:

There is great astronomical news. Mr. Piazzi, Astronomer to the King of the two Sicilies, at Palermo, discovered a new planet the beginning of this year, and was so covetous as to keep this delicious morsel to himself for six weeks; when he was punished for his illiberality by a fit of sickness, by which means he lost the track of it; and now a german Astronomer, having got some of his observations, has calculated its orbit in our system as near as he could from such few observations, and had just informed us where he thinks it should be looked for in the course of the summer and autumn.

It will not be so easy to recover, as the lost Cupid, when Venus said you might spy among 20 immediately by his air and complexion. But this having been only a star of the 8th magnitude at first, & now for some months to come not bigger than the 10th or 12th will not be easily distinguished among 40 000 or 50 000 stars of a similar appearance as it can be only known from them by its motion, which cannot be seen immediately but require observations of the relative position of several stars among which it is to be looked for.

What a deal this imprudent Astronomer has to answer for! It is now publicly proposed, in a german publication, to all the Astronomers in Europe to hunt for it.[3]

Zach published Piazzi's observations in the September 1801 issue of his *Monatliche Correspondenz*. . . (a publication Maskelyne later told Gauss that he could not obtain because of the war) and it was from this that Gauss obtained his data for the orbit. It was not until 22 October that Maskelyne eventually received a letter from Piazzi with full details.[4] In view of this astronomically exciting news, the Maskelyne family did not make its usual visit to Wiltshire that autumn.

In a letter to Banks dated 14 January 1802, Zach claimed priority for the re-discovery of Ceres, as mentioned in Maskelyne's account above. In fact, Maskelyne had already had the news from Méchain in Paris, and this had allowed him to find Ceres at Greenwich on 3 February and pass the position to Herschel, Aubert and Stephen Lee. (Preliminaries of peace with France began on 1 October 1801, the Treaty of Amiens was signed on 27 March 1802, and peace was proclaimed on 29 April.)

On 28 March 1802, Olbers at Bremen discovered a second new planet, which he named Pallas, and on 2 September 1804 Carl Ludwig Harding at Lilienthal discovered a third, Juno. All these proved to be very small bodies compared with the known planets and they came to be known as minor planets, or (named by Herschel) asteroids.

These discoveries meant a great deal of correspondence for the Astronomer Royal, who in those days had no secretary, every letter being written in ink in his own hand. The young German mathematician Carl Friedrich Gauss badly needed Maskelyne's observations (because they had such a reputation for accuracy compared with those made on the continent), as did the French astronomers. And they also meant the establishment of good relations with Zach, who had become *persona non grata* during the Mudge affair in 1793. Writing to Banks on 14 May, Maskelyne asked that he should 'assure him [Zach] that I have the highest esteem of his abilities & exertion in Astronomy, & think nothing of former disputes, & shall think myself honored by his correspondence.'[5] Soon after, he received a letter from Zach himself, bemoaning the fact that he could not measure zenith distances because Ramsden had failed to finish the promised 8-foot circle for Seeberg before he died (£200 had been advanced, but Zach had written this off, and ordered anew from Troughton), and also asking Maskelyne for his portrait so that he could have it engraved for his periodical; he said Chabert had taken a portrait to Paris (presumably an engraving of the Van der Puyl painting on p. 144) but Lalande declared it to be a very bad likeness.[6]

The Investigator

On 18 July 1801, Captain Matthew Flinders sailed in the *Investigator* to carry out a survey of the coasts of Australia. When Maskelyne had asked John Crosley to go on this important voyage as the Board of Longitude's observer – with chronometers, a clock by Earnshaw, and the usual outfit of instruments and equipment – he had agreed only if the accommodation was comfortable, saying that, in the *Providence*, his sleep had been disturbed and his health injured because his hammock was continually knocking against the partition: the *Investigator* was more roomy, he was assured.[7] However, Crosley fell ill on the first leg of the voyage and had to leave the ship at Capetown, having transferred the Board's instruments (except for two defective chronometers) to Flinders. As soon as the Board heard this, they made preparations to send out a replacement. The man chosen was a Cambridge mathematician, James Inman (whose *Nautical Tables*, first published in 1823, were until recently still a standard work in the Royal Navy), and among the instruments which he was to take was Kendall's third chronometer, to replace the defective ones brought home by Crosley.

Inman sailed with his instruments in the *Glatton*, arriving in Sydney about April 1803. He found that Flinders was at sea, so occupied himself in setting up an observatory on Garden Island, about a mile to the east of the spot Dawes had set up his observatory fifteen years earlier.[8] When Flinders reached Sydney in June 1803, it was found that the *Investigator* was in such bad condition that she could not continue with the survey. Flinders therefore decided to sail for home as a passenger in the old Spanish prize *Porpoise*. As there was no room in the *Porpoise* for the larger instruments, Inman was left behind in Sydney with instructions to find his own way home as best he could.

The *Porpoise* sailed with the *Cato* on 10 August. Then, on 8 September, Flinders himself arrived back in Sydney with the news that both ships had run aground on the Great Barrier Reef, he himself having made his way back in one of the ship's boats. New plans were made. Flinders wanted to get his precious documents home as soon as possible and elected to go himself in the tiny schooner *Cumberland*. The rest of the crew were to go in the *Rolla* to China where they could transfer to an East Indiaman for passage home.

Inman sailed from Sydney in the *Rolla* on 21 September 1803, reaching Wreck Reef a few days later where she embarked the remainder of the crews of the *Porpoise* and *Cato*, who had been camping on the reef since the disaster seven weeks before. Flinders recovered the Board's instruments from the *Porpoise* and turned the majority of them over to Inman. On 11 October, the *Cumberland* sailed for England with Flinders, and the *Rolla* sailed for Canton with Inman and the bulk of the *Investigator's* ship's company. Flinders found it necessary to put in to the Ile de France (Mauritius), but reckoned the documents he carried would give him safe conduct despite the state of war. But the French Governor thought otherwise and, to his chagrin, Flinders was kept prisoner on the island for seven long years. Inman joined the *Warley* East Indiaman at Canton, having an eventful voyage home because the East India Company fleet under Sir Nathaniel Dance was involved in the celebrated engagement when the ships of the French Admiral Linois were driven off.

There was no lack of excitement for observers appointed by the Board of Longitude!

Inman eventually returned the chronometers to Maskelyne at Greenwich on 16 August 1804. (After being ordained, Inman became Professor of Mathematics at the newly-constituted Royal Naval College at Portsmouth in 1808, when William Bayly retired from the post of headmaster of the erstwhile Royal Naval Academy.)

The Earnshaw affair

In his memorandum book under July 1789, Maskelyne mentioned for the first time someone who was to cause him a great deal of work – and not a little anxiety – during his declining years:

Mr. Hernshaw, Watchmaker, No. 28 Shaftsbury Place, Aldersgate, says he invented the Detached escapement used by Arnold in his watches.[9]

He was actually referring to the 40-year-old Thomas Earnshaw whom a mutual friend had brought to the observatory on 1 July, with a watch which his visitor claimed was much better, and far cheaper, than any made by Arnold or other chronometer makers of the time.

Three days before Earnshaw's visit, Maskelyne had started the third trial of Mudge's chronometers 'Blue' and 'Green', together with a box chronometer by Arnold, so he placed Earnshaw's watch alongside them in the Transit Room and began to make daily comparisons with the transit clock. At the next meeting of the Board of Longitude, on 15 August, he reported that in a six-week trial 'Mr. Hernshaw's' watch had shown great promise, and he was directed to continue the trial.[10] According to Earnshaw himself, Maskelyne then suggested he should start making two identical chronometers to offer for trial in accordance with the Longitude Act of 1774.

Maskelyne was much impressed by Earnshaw's workmanship and it was through his influence that the latter received orders from the Board of Longitude and others for several chronometers and clocks in 1791. He made extensive alterations to Graham's transit clock at Greenwich in 1793, and had another of his watches tried there in 1796–7.

Eventually, the two chronometers specially made to qualify for a Longitude prize went on trial at the beginning of 1798. Over the next five years, they were given three separate one-year trials at Greenwich, and hardly a Board meeting went by without a petition from Earnshaw asking for money or complaining about the conduct of the trials. Only once was he successful when, in December 1800 after the second trial, he was advanced £500 'to encourage his endeavour to obtain the Parliamentary reward'. Maskelyne was closely involved in all these exchanges.

John Arnold died in 1799, his business being continued by his son, John Roger Arnold.

The third Earnshaw trial ended in July 1802, so, for the December Board meeting, Earnshaw presented a petition claiming the results were such that a sea trial should now be ordered, as the Act demanded. But no, said the Board, by our arithmetic your chronometers have not gone within the limits prescribed by the Act, though they heard evidence from Matthew Flinders in Australia that Earnshaw's chronometers had gone better in the *Investigator* than Arnold's.

With the greatest reluctance, Earnshaw accepted the decision that he did not qualify for the highest awards, but wrote to the Board claiming a lesser reward, at least equal to the highest any previous person had received under the 1774 Act, the £3000 received by Mudge in 1793. On 3 March 1803, the Board met at the Admiralty as usual. Admiral Earl St Vincent was in the chair, with Banks, Maskelyne, Professors Hornsby and Robertson from Oxford, and three from Cambridge whom we have not met before, Isaac Milner, Samuel Vince, and William Lax. Deciding that Earnshaw's chronometers had indeed gone better than any of those previously submitted for trial at Greenwich, they resolved – unanimously according to Maskelyne[11] – that he should be awarded £2500 (to be added to the £500) as soon as he had disclosed his secrets. Banks was desired to wait upon the Prime Minister (Henry Addington) to ask for an additional grant for the Board. A few days later, Maskelyne and Robertson called on Earnshaw at his house in High Holborn to tell him the good news.

Then began the first rumblings of troubles to come. Shortly after the visit to Earnshaw, St Vincent called an extraordinary meeting of the Board for March 17 'in consequence of doubts having arisen relative to the accuracy of the assertion in the Minute of the last Board which declares Mr. Earnshaw's time-keepers to have gone better than others that have hitherto been tried at the Royal Observatory.' In view of subsequent events, one assumes it was Banks who raised these doubts, on behalf of his erstwhile friend, the late John Arnold. However, at the extraordinary meeting, Maskelyne produced evidence refuting the claim that Arnold's chronometers had performed better than Earnshaw's in Greenwich trials. The previous minute was confirmed, again unanimously.

Now a joker appeared in the pack. From the very beginning, Earnshaw claimed that it was he who had invented the improved detached escapement which Arnold had patented in 1782. What if Earnshaw's claim was not true? Also, he was using a compensated balance and balance spring said by some to have been invented by Arnold. The Board meetings in June and December 1803 and on 1 March 1804 were largely taken up in hearing evidence on these matters from other chronometer makers. Then, on 19 March 1804, there appeared a printed pamphlet: *Sir Joseph Banks's Protest against a Vote of the Board of Longitude, granting to Mr. Earnshaw a Reward for the Merit of his Time-keepers*. In it, Banks made the following points:

(1) In his opinion, the Act did not limit consideration to chronometers tried at Greenwich;
(2) more rigorous trials were needed; and
(3) Earnshaw was using improvements invented by Arnold.

With the assistance of George Gilpin, Secretary to the Board and sometime Maskelyne's assistant at Greenwich, he published tables which he said proved the superiority of Arnold over Earnshaw.

At another extraordinary meeting on 23 March, Robertson suggested that the Board might be better able to form reasonable conclusions if drawings and models of both makers' escapements, at five times full size, could be provided. Earnshaw and John Roger Arnold duly produced the models in time for the June meeting. Then, just before the December meeting, Maskelyne entered the fray once more, publishing a 14-page pamphlet: *Arguments for giving a Reward to Mr. Earnshaw...* This was a reply to Banks's *Protest*, Maskelyne's concluding paragraph giving a good summary of the position at the time:

Now, I think, I have absolutely shewn that Earnshaw's two watches, made a few years ago, have gone better than the picked and veteran watches of Arnold and Emery; and that Earnshaw, if he has not absolutely shewn to be the original and sole inventor of the detached Escapement, yet, at least, he has shewn himself to be an independent and contemporary inventor with Arnold, and as such entitled to a reward from this Board for so highly useful an invention, which Arnold never claimed before this Board, or solicited a Reward for; and which is now open to Earnshaw; who has also a claim to a further reward for his other improvements in Time-keepers, if the Board will accept his communication of the whole construction and modes of adjustment of this Time-keepers.[12]

Meanwhile, Pitt had replaced Addington as Prime Minister, appointing a new First Lord of the Admiralty (who knew something of the problem, having been Treasurer of the Navy during the Mudge affair). The Board of 6 December 1804 must have been a formidable affair. Earnshaw (who was not present) describes the scene:

At this meeting Lord Viscount Melville, First Lord of the Admiralty, sat in the chair; the Astronomer Royal, the Professors, and the rest of the Board of Longitude [actually, only the Comptroller of the Navy] were in my favour; Sir Joseph, in a single minority of himself, sheltering himself behind the main-sail of the new Act of Parliament, against the abovenamed gentlemen, till Lord Melville made the following proposition to bring Sir Joseph to, which I am informed by some who were at the Board, his Lordship made in the following words, 'That he found Sir Joseph Banks on one side for Arnold; the Astronomer Royal and the Professors for Mr. Earnshaw, that he could not differ with those gentlemen, as they certainly knew most of the matter, that he himself had the highest opinion of Earnshaw, but he did not like to differ with Sir Joseph; and in order to reconcile both parties, proposed to make Arnold's reward equal with Earnshaw's.'[13]

The minutes merely record this Alice-in-Wonderland-like decision, telling nothing of the drama which must have preceded it. The money was not to be paid until both parties had disclosed on oath full explanations in writing. This was done and copies were sent to a dozen or more chronometer makers for comment.

An extraordinary meeting was called for 11 July specially to consider these comments. However, when Maskelyne and the professors arrived at the Admiralty, they were greeted with a message from the Lord Barham, the new First Lord (Melville had resigned in May in the wake of a scandal about naval funds), saying that, as Banks could not attend, could the Earnshaw–Arnold business be postponed until December?

Exactly what Banks hoped to achieve from these delaying tactics we do not know, but whatever it was seems to have failed because he pointedly absented himself from the meeting on 12 December 1805 when the Board authorized the payment of £2500 to Earnshaw and £1678 to Arnold which, with the sums already granted, brought the rewards up to £3000 each.

Everyone hove a sigh of relief: that was hopefully that. But no! On 4 February 1806, there appeared in the *Morning Chronicle* (later repeated in the *Times*, *Morning Post*, *Sun*, *Star* and *Courier*) a long advertisement by Earnshaw giving his version of events, full of invective against 'Sir Joseph Banks, and Mr. Gilpin, Secretary to the Board, [who] used extraordinary exertions against Mr. Earnshaw, and encouraged the watchmakers against him', accusing Banks of being the head of a combination of wicked and malicious men conspiring together to opress an individual.

Banks was furious, writing to Maskelyne to say he did not intend to remain quiet under these accusations.[14] In reply Maskelyne said that, while the professors considered the advertisement most improper, they supposed he would not condescend to answer it.[15]

Having obtained Counsel's opinion that the advertisement was indeed libellous,[16] Banks demanded at its meeting on 6 March that the Board (chaired by yet another First Lord, Charles Gray) should prosecute Earnshaw for reflecting on his character while in the exercise of his duty to the Board. However, the Board, while affirming that it had the highest opinion of Banks's labours on its behalf for the last 25 years and disapproved strongly of the advertisement, did not think it expedient to order a prosecution. In that case, said an indignant Banks, I will begin a prosecution privately. In the event he did not do so, so future historians have been denied the pleasure of reading about these two strong personalities facing each other in a court of law.

That was the last meeting of the Board of Longitude that Banks attended until after Maskelyne's death in 1811. In 1818, he achieved his desire when the Board was reconstituted so as to increase the power of the Royal Society at the expense of 'the professors'.

The rest of the story can be quickly told. In March 1806, Alexander Dalrymple, Hydrographer to the East India Company and now Hydrographer of the Navy, a friend and patron of Arnold senior, published an 88-page *A Full Answer to the Advertisement...*,[17] refuting Maskelyne's *Arguments...* and giving examples of Arnold's excellence. This was answered in turn in February 1808 by Earnshaw's 313-page 'pamphlet' *An Appeal to the Public...*,[18] full of invective in which even the professors did not come out unscathed. He also castigated the anonymous correspondent who had ascribed all of Earnshaw's inventions to Arnold in the August 1806 issue of Nicholson's Journal.

The same month, Earnshaw presented a petition to Parliament, suggesting that, as his chronometers were so much better than Arnold's, he should have a far greater reward. The petition was ignored that year, but, in February 1809, he submitted it again. This time, a parliamentary committee took evidence from many watchmakers and Board members, and then recommended that Parliament should not interfere with the Board of Longitude's decision. (Despite their differences, Maskelyne wrote to Banks describing his own examination.[19])

So ended the Earnshaw affair, though who invented the modern detached escapement is still a subject of debate among horological historians – and probably always will be!

Personal and observatory affairs

Though in 1801 in his 70th year, Nevil Maskelyne did not allow his age to stop him from living a very busy life. This letter to his sixteen-year-old daughter (addressed to Miss Maskelyne, Royal Observatory, Greenwich) gives some flavour of this busy life:

> Soho Square, Friday Morning
> 10 o'clock Novr. 20 1801

My Dear Margaret,

I was apprised last night by Baron Maseres, who came on purpose to the Royal Society, that the [Friday] club would meet today [Maskelyne had dined at the Royal Society Club the previous evening]; so I shall sleep at Wright's [Coffee House] tonight. I intend to come home to the Observatory tomorrow morning; but if any thing should happen to prevent me, I desire you will take me up here

Fig. 15.1. *Nevil and Sophia Maskelyne*, 1801 & 1803, miniatures, both signed M. Byrne, in the possession of Nigel Arnold-Forster, Esq. Maskelyne paid Mary Byrne £10 for each, including frames and gold mounts, in November 1801 and July 1803. In her portrait, Sophia is wearing the miniature of Nevil. (Photo: author)

Fig. 15.2. *Nevil Maskelyne*, 1801, miniature probably by Theed jr., whose father William Theed was paid 3 guineas for a miniature on 22 December 1801. In the possession of Nigel Arnold-Forster Esq. (Photo: author)

at six o'clock at furthest. I have sent Sam with a note to Mrs. Talmash, Newcastle Place, to provide the two beds for us. Inclosed is a note for Mr. Walker about the coach, which had best be sent this afternoon. Bring the Play book with you. Be sure you bring Madame Gouvernante with you. I saw Mendoza [the Spanish astronomer] & the Marquis de Chabert at the Royal Society.

Yesterday I provided myself with a pair of lamb's wool socks and immediately put them on to wear when I walk in dirty streets.

It was thro' the forgetfulness & neglect of the people at the globe [tavern] that they sent no notice about the meeting of the club last week.

The Royal Society meeting was extraordinary thin; but whether owing to the coldness of the weather, or the attractions of Mrs. Billington [a soprano who was attracting rave reviews for her performance in Sheridan's *The Duenna* at Drury Lane[20]], I am in doubt; but perhaps on second thoughts, partly to one, and partly to the other.

My love to your mama & yourself
from,

    Dear Margaret,

        Your affectionate Papa
        N. Maskelyne[21]

One feature of these later years was that, when the Astronomer Royal took his holidays, it was generally for a longer period than previously, perhaps because he now had an Assistant he could trust. His Account Book records, for example, that in 1800, he was absent in Wiltshire from 16 September to 30 November. He took no holiday in 1801, presumably because of the Ceres discovery, nor is there any record of one in 1804. His last visit to Wiltshire seems to have been from September to November 1806. In 1809, he took the family for a month to Ramsgate, where Samuel Vince and his wife were staying.

In his accounts for 1800 we find the first mention of Income Tax – a payment of £124.3.2½ due on December 5.[22] Pitt's tax introduced in 1799 was at the rate of 10%, but it is possible that the tax paid here covered a period of more than a year. His official emoluments were still only £300 a year, but he had a considerable private income.

Maskelyne had received many foreign honours during his professional career – from Hanover, Russia, Poland, and the new United States of America – but the honour he probably appreciated most was his election on 24 February 1802, just before the Peace of Amiens, to be one of only eight foreign associates of the newly-founded *Institut National des Sciences et des Arts* in Paris, despite the fact that, with the other two English associates, Banks and Cavendish, he was severely criticised in Britain for having accepted the honour. He received the silver medal that went with it on 6 June 1804, by which time Britain was once more at war with France.

The amount of work caused by the Board's publications grew no less. This was a time of considerable monetary inflation and in March 1802 Maskelyne was able to persuade the Board to increase the salary of the

Fig. 15.3. *Nevil Maskelyne*, 1804, aged 72, pastel by John Russell RA, in the possession of Nigel Arnold-Forster, Esq. John Russell painted this pastel and its pair in gratitude for Maskelyne's help in producing his moon pastels and his Selenographia, which was 'A Globe representing the Visible Surface of the Moon, constructed from Triangles measured with a Micrometer and accurately drawn & engraved from a long series of telescopic Observations by J. Russell, R.A.', published in 1797. 'Nobody is pleased with your father's little portrait' was Lady Clive's comment to Margaret on 1 July 1804. (Photo: NMM)

Fig. 15.4. *Sophia Maskelyne*, 1804, aged 52, pastel by John Russell, RA; the pair to Fig. 15.3. (Photo: NMM)

computers of the Nautical Almanac from £140 to £180 per almanac, and also to give the comparer, the Rev. Malachy Hitchins, an augmented rate of £220. Before he died, Wales had prepared a third edition of the Requisite Tables and this was eventually published on 10 December 1802. A little earlier, the opportunity had been taken to reduce stocks of some of the Board's unsold publications, the complete stock of 6000 copies of the first edition of the Requisite Tables (10 000 had been printed in 1766), 674 copies of Mayer's Tables (keeping 500; 2000 had been printed in 1770), and many old almanacs being sent to be pulped in December 1800.

In Maskelyne's Memorandum Book of 1804, sandwiched between a note that he had sent copies of Earnshaw's Explanation to Professors at Oxford and Cambridge and an extract from *The Times* about 'the symptoms, medicines, regimen, & method of treating yellow fever', is the following entry:

Nov.16 – found an alteration in my speech, something of a paralytic affection.[23]

He was 72 when that was written. The following year, on Whit Monday, 3 June 1805, he received another royal visit to the observatory. At noon, the King, accompanied by the Prince of Wales, Duke of York, and Duke of Cambridge, reviewed the Kent Volunteers on Blackheath, after which they went to the residence of Caroline, Princess of Wales (already formally separated from the prince) in Montague House in the southwest corner of Greenwich Park (it was demolished in 1815) where they were joined by Queen Charlotte and five princesses. *The Times* continues the account:

The King, after the review was over, was escorted to the princess of Wales's, where her Majesty and the Princesses arrived about two o'clock. The whole party, except the Duke of Cambridge, sat down to an elegant dinner; after which, they went in the Princess of Wales's open carriages to Flamstead House, and viewed the Royal Observatory. They were conducted by the Astronomer Royal.

They then proceeded to view the holiday-makers' gambols of rolling down Greenwich Hill, and other sports of the Fair, with which they appeared very amused.[24] (Fig.15.5)

At the annual Visitation a month later (the last to be attended by the Astronomer Royal's great friend, Alexander Aubert, who died in October), Maskelyne once again raised the matter of the provision of a new circular instrument. A year later, in 1806, he submitted to the Visitors a formal memorial, pointing out that, as a result of the great improvements made in the construction of astronomical instruments in the last twenty years (principally by British instrument makers), most foreign observatories were now furnished, not with quadrants, but with divided circles for observing the distances of celestial objects from the zenith, capable of far more accurate observations than his own mural quadrants. Furthermore, Bird's 56-year-old brass mural quadrant (Graham's iron quadrant was 81)

The final years, 1800–11

Fig. 15.5. *Greenwich Park on Easter Monday*, from an engraving by Pass after Pugh, 1804. After the King and Queen had visited the observatory on Whit Monday, 1805, they 'proceeded to view the holiday-makers' gambols of rolling down Greenwich Hill, and other sport of the Fair'. Writing of Greenwich Fair some years later, Charles Dickens said that in the park 'the principal amusement is to drag young ladies up the steep hill which leads to the Observatory, and then drag them down again, at the very top of their speed, greatly to the dearrangement of their curls and bonnet-caps and much to the edification of the lookers-on below.' (Photo: NMM)

now gave a different latitude from that found by Bradley – a sure sign of wear on the pivots.

He suggested that the Board of Ordnance be asked to provide a circular instrument of dimensions as suggested by him, together with a building to accommodate it, and a clock to go with it.[25] The hope was that this one instrument would be capable of measuring both zenith distance and right ascension, thus replacing both the two mural quadrants and the transit instrument. (In the event, this proved not possible to the precision needed, so a new transit instrument was mounted in 1816.)

A month earlier, on 26 June 1806, the 39-year-old astronomer John Pond had communicated a paper to the Royal Society giving the results of observations made with his 30-inch altazimuth circle by Edward

Troughton, mounted by his house in Westbury-sub-Mendip, Somerset. These observations gave decisive proof of the deformation of Bird's quadrant at Greenwich.[26] Pond had impressed Maskelyne with his skill as an observer and it must have been at this time that the latter began to realise that here perhaps might be a suitable successor as Astronomer Royal. Maskelyne noted that Pond dined with him at the observatory on 8 and 16 August. (A fellow-guest on the 16th was Dr Charles Burney, brother to Fanny the diarist, who had a school at Greenwich.)[27]

Pond was elected FRS in February 1807, marrying and moving to London the same year. He is said to have spent much time in the workshop of his friend Edward Troughton, who had begun work on the Greenwich 6-foot mural circle, ordered after he had shown a model of his proposed design to the Visitors on 28 May 1807. Pond gained a deeper knowledge of the work of the observatory when Maskelyne gave him the task of revising for the Press the Greenwich Observations for 1806 – for which he was paid 2 guineas.[28]

Meanwhile, on behalf of the Bureau des Longitudes in Paris, Delambre had, in 1806, sent Maskelyne seven new tables of the Moon and sun compiled by Bürg and himself respectively, derived from 'the greatest and most valuable compilation of observations in existence' – in other words, Maskelyne's own. 'Kindly accept therefore,' continued Delambre, 'a work to which you have made such a major contribution. We shall be very flattered if you judge our tables worthy to be used in the calculation of the Nautical Almanac, in accordance with the hope given to us in your latest preface.'[29] The Nautical Almanac computers found the new French tables easier to work with, even though more work was involved. They were used for the 1813 and subsequent almanacs.

On 1 July 1807, Thomas Taylor replaced Thomas Ferminger as Assistant. (Taylor remained at Greenwich with Pond, becoming First Assistant in 1816, a post he continued to hold until dismissed when Airy succeeded Pond in 1835.)

The new mural circle was to be erected on a massive wall 4 feet thick and 6 feet high in the centre of a new Circle Room to be built onto the east end of Bradley's New Observatory. William Hardy, who was making the new clock to replace Graham's transit clock of 1750, had this to say in 1820:

At that time they were making great improvements at the Observatory which engrossed all the Dr's attention. The Room for the new Instruments was then building, and such was his perseverance that he was out there in the most inclement weather superintending the workmen, which I believe tended much to shorten his days. . .[30]

One wonders how the workmen viewed this superintendence!

The death on 28 March 1809 of Rev. Malachy Hitchens was a great blow. For forty-three years, mostly working in Cornwall, he had been the 'comparer' of the two independent computations of every table in every almanac and was the linchpin of the organization. Now, as a temporary measure, Maskelyne arranged that the more senior computers – Henry Andrews, Nicholas Games, and Mrs Mary Edwards – should share the comparing work. Nevertheless, as he told the Board in December, an enormous extra burden was thrown upon himself. Earlier that year, the computers had applied for an increase 'in consequence of an advance of price of every article of life.'[31] The Board increased the allowance per almanac from £180 to £225 for computers and £220 to £250 for comparers. For the first almanac in 1766, the rate had been only £70.

Even Maskelyne himself felt the effects of inflation. Some time late in 1809, he drafted a memorial to the Lords of the Treasury asking for a rise in salary for himself and for his Assistant. He pointed out that the Astronomer Royal's total emoluments – a gross sum of £350 a year, reduced by fees, etc., to some £300 – had not altered since 1752 and those of the Assistant – only £96 a year – had remained the same since 1771. He asked, 'That, on account of the alteration of the times, and the increased price of every thing, all these emoluments are now very inadequate to support himself and his Assistant, and therefore he humbly requests your Lordships to take the same into your consideration, and represent it to his Majesty, whose goodness and liberal regard for science, will, he doubts not, incline him to grant an augmentation to his appointments. . .'[32] Scored out in the draft that survives is the suggestion that such an augmentation should not be less than £350 a year for himself and £100 for the Assistant.

Banks went to see Spencer Perceval, Chancellor of the Exchequer and Prime Minister. His description of the interview in a letter to Maskelyne is worth quoting in full:

Soho Square
Jan.8. 1810.

My dear Sir

I have seen Mr. Percival on the subject of your request to have the emoluments of your Office & that of your Assistant increased, but I can not say I find him so propitious as I hoped to have done.

He admitted the depreciation of the value of Money since your Salary was fixed, but when he saw your proposal of doubling the whole amount of yours &

your Assistants, he startled much, for how evident soever it may be that Money will not now go half so far in providing the necessaries of life as it did 40 years ago, & that the Emoluments of many Functionaries have on that account been increased, none that I am now aware of have been actually doubled.

He enquired concerning your circumstances & being told that you possessed besides your appointment a comfortable patrimony, he seemed to think that he would find it an easier matter to settle the quantum of increase (for he admitted that increase would be proper) with your successor than with you, to whom personally it could not be any great object, but he ended by saying, now do not you think Sir Joseph that there are able Men both of Oxford & Cambridge who would be happy to undertake the business of the Royal Observatory for £300 a year & a good House? to which I did not venture to deny my assent.

On the subject of the Assistant he seemed more inclined to be gracious, but not in any degree inclined to double his present pay, when we meet we will talk over these matters again & consider what is best to be done.

believe me my dear sir,
Your faithfull & hble Servt
Jos: Banks.[33]

... which hardly shows an objective approach!

As a result, Maskelyne drafted a new memorial which was forwarded by the Royal Society Council to the Treasury on 29 March 1810. In this he omitted all mention of his own salary but asked for an augmentation of £100 for that of his Assistant, pointing out that often, as soon as an Assistant had been trained up, he left simply because of the low pay, 'to the great inconvenience of your Memorialist and to the obvious detriment of the service of the Institution.'[34]

On 1 May 1810, the King put a very shaky signature on a Warrant granting the Assistant this augmentation from the beginning of the year.[35] The following year – 1811 – Spencer Perceval authorized a salary of £600 a year for the new Astronomer Royal.[36] The year after that – 1812 – Spencer Perceval was assassinated.

The final months

Nevil Maskelyne's last year was as busy as ever and he had a large amount of paper work to deal with, particularly on Nautical Almanac business. Presumably out of pique, Banks had not attended a meeting of the Board of Longitude since the Earnshaw affair in 1806, so Maskelyne generally took the chair except on the few occasions when the First Lord was present. He attended all three meetings in 1810, taking the chair at his last, on 6 December, when he welcomed the new Savilian Professor of Geometry at Oxford, S. P. Rigaud, replacing Robertson who had transferred to the chair of Astronomy in place of Hornsby who had died earlier in the year.

Maskelyne was a regular attender too at the Royal Society and its Council, his last Council being on 13 December 1810. During that year, he attended ten dinners of the Royal Society Club, at the last of which, on 23 August, he sat next to James Smithson, the future founder of the Smithsonian Institution in Washington, D.C., even though a citizen of Great Britain.

At the annual Visitation at Greenwich on 3 August, Maskelyne was able to show the Visitors – led by Banks, supported by Humphrey Davy (Vice President), John Pond and six others – the completed Circle Room, but, alas, not the mural circle itself which, despite being signed *Troughton London 1810*, was not erected at Greenwich until June 1812, too late for him to see the fundamental instrument he had so long desired.

At noon on 1 September 1810, he took what proved to be his last astronomical observation – a transit of the Sun. From about this date, the handwriting in his letters and account book suddenly deteriorated though there was no deterioration in his mental capacity. On 25 October, 'being in indifferent health but (thanks be to God) of sound and disposing mind, memory and understanding,' he made his will.[37] He was, however, well enough to attend meetings of the Board and the Society in December.

The last two of his letters which have survived were both written on 5 January 1811, only a month before his death; both are perfectly lucid if in a fairly shaky hand – one to Henry Andrews, highly mathematical on almanac matters;[38] the other to Vince staying at Ramsgate (illustrated in Fig.15.6), asking him and his wife to visit the observatory on the way to the next Board of Longitude (on March 7), and answering a geometrical query.[39] Vince told Margaret later that he was much grieved to see by the handwriting how ill her father was, 'and yet', he added, 'the Doctor did the problem right.'[40]

On 7 February, Margaret wrote to Henry Andrews, 'I write at Dr Maskelyne's desire to inform you that he is very ill, and consequently the business of the Nautical Almanack must stand still for the present.'[41] Two days later, on Saturday, 9 February 1811, after an illness of only three weeks in which even then he did not take to his bed, he died at the Royal Observatory in the 79th year of his age, having completed 46 years as Astronomer Royal.[42]

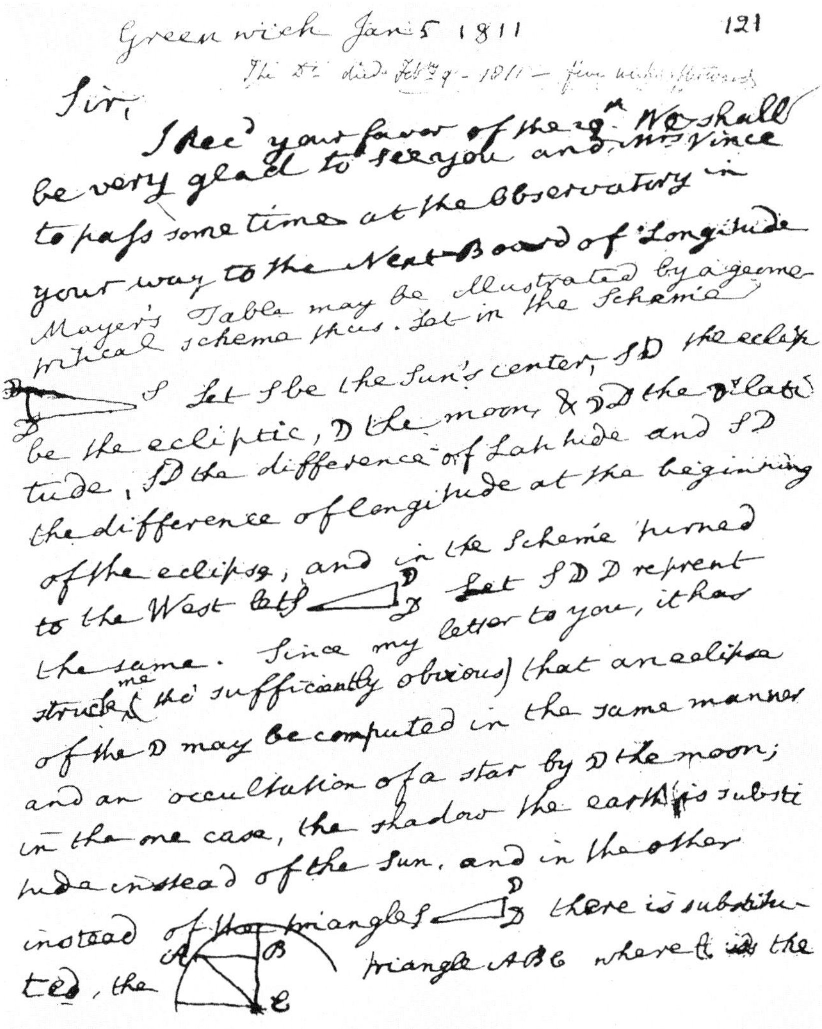

> Greenwich Jan 5 1811   121
>
> *The Dr. did Feb 9 - 1811 - five weeks afterwards*
>
> Sir,
> I rec'd your favor of the 2d. We shall be very glad to see you and Mrs Vince to pass some time at the Observatory in your way to the Next Board of Longitude. Mayer's Table may be illustrated by a geometrical scheme thus. Let in the Scheme
> Let S be the Sun's center, SD the eclip be the ecliptic, D the moon, & DD the D's latitude, SD the difference of Latitude and SD the difference of longitude at the beginning of the eclipse, and in the Scheme turned to the West &c Let SDD reprefent the same. Since my letter to you, it has struck me (tho' sufficiently obvious) that an eclipse of the D may be computed in the same manner; and an occultation of a star by D the moon; in the one case, the shadow the earth is substituted instead of the Sun, and in the other instead of the triangles there is substituted, the triangle ABC where B is the

Fig. 15.6. The Last Letter Written by Nevil Maskelyne – to Samuel Vince at Ramsgate, 5 January 1811, five weeks before his death. Vince told Margaret Maskelyne later that he was much grieved to see by the handwriting how ill her father was, 'and yet,' he added, 'the Doctor did the problem right.' (NMM MS PST/76, 119–21. Original in possession of Nigel Arnold-Forster, Esq.)

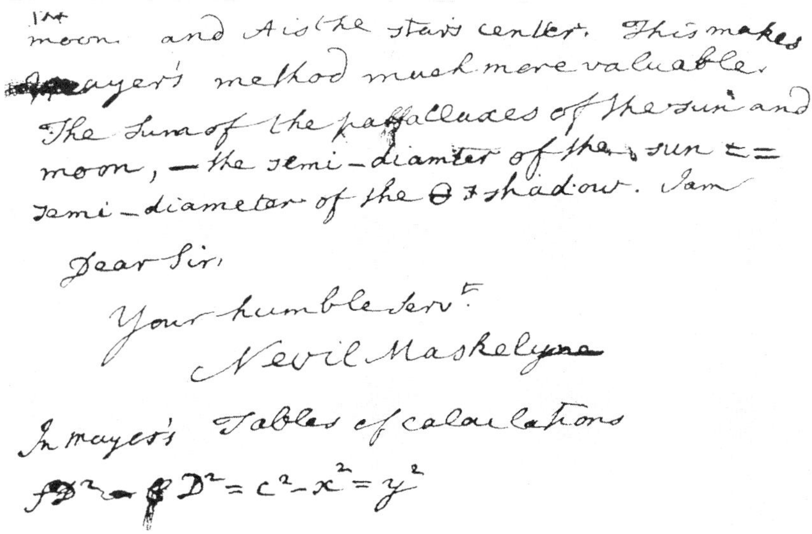

[Handwritten letter excerpt:]

moon. and A is the stars center. This makes Mayer's method much more valuable. The Sum of the parallaxes of the sun and moon, — the semi-diameter of the sun ± = semi-diameter of the ☉ ☽ shadow. I am

Dear Sir,
    Your humble servᵗ
        Nevil Maskelyne

In Mayer's Tables of calculations
$f\mathcal{D}^2 - f D^2 = c^2 - x^2 = y^2$

Epilogue

Nevil Maskelyne was buried at the church of St Mary, Purton, Wiltshire, on 20 February. His tomb is near the south wall, not far from his memorial in St Nicholas Chapel, the text of which is given in Appendix F.

In his will, he bequeathed Purton Down Estate to his widow Sophia, the rest of his estates (including Basset Down) to his daughter Margaret without any entail (she was 25), his personal effects, books, etc., to Sophia, and the invested funds equally between them. He set up trusts to provide £30 annually to James Houblon Maskelyne's widow and daughter, both called Ann; and £52.10s annually to William Maskelyne's putative daughter, Mrs Jane Sacheverell. He made the following bequests: Margaret Dowager Lady Clive (sister) £100; Frances Margaretta and Charlotte Walpole, and Charlotte and Harriet Robinson (great nieces on the Clive side) £50 each; Joseph Toomer (relict of James Houblon Maskelyne's other daughter Jane) £300 in 3% consols, which, with duty, produced £175 net; Rev. Samuel Vince £50; Dr Charles Hutton £20; Richard Carter (cousin) £20; the British Lying In Hospital for Married Women in Brownlow Street, Long Acre, London, £20.[43]

He was succeeded as Astronomer Royal by John Pond, his own choice, obviously approved of by Sir Joseph Banks, and recommended to the Prince Regent by Humphrey Davy, who had visited him at Westbury in 1800.[44] Pond's first observation at the Royal Observatory was on

Fig. 15.7. The Church of St Mary, Purton, Wiltshire, where Nevil Maskelyne was buried on 20 February 1811. See Appendix F. (Photo: author)

11 January, some time before the death of his predecessor: he took possession of the observatory and its contents on 13 April.[45]

Nevil Maskelyne's library was sold by Leigh and Sotheby on 27 May 1811, a three-day sale of 757 lots which realised a total of £451 18s.6d,[46] from which Margaret Maskelyne received £369 13s.2d., plus £26 5s for the camera obscura. In the vitriolic article about Banks already mentioned, Olinthus Gregory tells how Mrs Maskelyne offered the library to the Government for Pond on a fair valuation, and how Banks persuaded the Visitors to decline. When the library was sold at auction, says Gregory, agents employed by Sir Joseph snapped up those books which he thought most valuable, implying that the widow was thereby swindled out of her just dues.[47] No confirmation of these scurrilous allegations has been found.

Margaret and her mother went to live at Basset Down. In 1819, Margaret married Anthony Mervyn Story, who subsequently took the additional name of Maskelyne. Their eldest son, Nevil Story Maskelyne (1823–1911) became a distinguished mineralogist. Sophia died in 1821 and is buried in the same grave as her husband, as were Margaret and her husband.

Memorandum book extracts

– Mr. Schwep, Mortimer Street, Cavendish Square, sells Soda water, full of fixt air. Major Campbell makes great use of it. Aug.1.1801. Mr Hulse

(MB.2H/37^v)

– 1801 Aug.13   Sir Joseph Banks takes now 3 drams of ginger in milk, at his breakfast. Thinks it may be taken in a bolus, & will not disagree with the stomach.

(MB.2H/37^v)

– To make ginger tea.
Take a quarter of an ounce of ginger, sliced very small, & put it in a tea-pot, and pour a pint of water over it, & let it stand terminates. – Take a large coffee cup of it, first putting a tea-spoonful of brandy in it. It is good against wind in the stomach, and gout in the stomach.
Mr. Wegg. Nov.20 1801

(MB.2H/39^v)

– Sir J. Banks will be 58 May 1802, born 1744.
– Sir Josh Banks abstains from wine, & drinks moderately strong brandy & water.
– Nov.19 1801. Bought a pr of lambs wool socks 2s. No. 28 in the Strand.

(MB.2H/40)

– 1805 Dec.5 – £178.18.4¼ collected at Greenwich Church, on the thanksgiving day, for the relations of the killed & wounded men in the sea-fight with the combined fleet off Trefalgar, on Oct.21st. The Princess of Wales sent £5 which makes it £183.18.4½.(*sic*)

(MB.2J/34^v)

– Places of Diversion
1806 Feb.26 – Oratorio
    May 3 – Belle's Stratagem & 40 thieves Dr. L
        6 – Barbarossa   ″
        8 – Astley's trial
        9 – Panorama of Bay of Naples & Sadler's Wells
     10 – Exhibition
        8 – Trial of Ld. Melville

July 16 – to Vauxhall
24 & 25 – To Oatlands & Shepperton. Panorama of Trefalgar.
West's picture of death of Nelson.
1807 Jan. 1 – To Covent Garden theatre
Feb.20 – Covt. Garden theatre. Acis & Galatea
June 20 – To the Opera Semiramide. Madame Catalani

(MB.2K/5)

16

Summing up

> His plans were mostly directed to substantial objects, while a steady perseverance gave an efficiency to all his undertakings: and notwithstanding his profound knowledge of physical astronomy, his attention was chiefly directed to reduce the scientific theories of his predecessors to the practical purpose of life. In this he was eminently successful, particularly in his labours for the longitude, by which he essentially contributed to the advancement of navigation, the prosperity of commerce, and the wealth, honour, and power of his country.[1]

This flowery quotation is from a biographical note on Nevil Maskelyne written shortly after his death, which continues:

> Thus, from Dr. Maskelyne's important labours, his public character is well known, and his fame immoveably established: and, as to his private character, it was likewise truly estimable. He was, indeed, exemplary in the discharge of every duty. In his manners, he was modest, simple, and unaffected. To strangers, he appeared distant, or rather diffident; but among his friends he was cheerful, unreserved, and occasionally convivial. . .
>
> Notwithstanding the doctor's numerous avocations he found time to maintain a regular correspondence with the principal astronomers of Europe. He was also visited by many illustrious foreigners, as well as eminent characters of his own country, but his warmest attachments were always manifested to the lovers of astronomy. Among his most intimate friends may be reckoned Dr. Herschel, Dr. Hutton, Messrs. Wollastons, Mr. Aubert, bishop Horsley, sir George Shuckburgh, baron Maseres, professor Robertson; and also professor Vince, whose publications so ably illustrate Dr. Maskelyne's labours, and whom he appointed the depositary of his scientific papers.[2]

Of course, such an elegy must be expected to contain a certain amount of hyperbole. Nevertheless, evidence from the large body of correspondence that survives proves that, pompous and a bit of a bore though he might have seemed to some, Maskelyne was almost universally liked and admired by his contemporaries – except perhaps by some chronometer makers and their families. It is perhaps significant that the list of intimate friends given above contains so many mathematicians. Despite frequent disagreements in committee, his personal relations with Sir Joseph Banks were more cordial than might be expected and, in society, they seem to have enjoyed each other's company and met as friends and intimates.

The reputation that survives in some popular books today, of Nevil Maskelyne as the evil genius who tried to deprive the poor illiterate Yorkshire carpenter, John Harrison, of his just rewards – out of personal spite and because of his, Maskelyne's, own involvement in the rival lunar-distance method of finding longitude – was certainly not one that was held generally in his own day, nor is it in any way justified by today's research. He was a member of the Board of Longitude, appointed by Parliament to advise in the award of large sums of public money. There is no evidence whatsoever that Maskelyne at any time abused his position as a public servant in order to further his own ends, still less to line his own pocket.

One of his more likeable traits was the way in which he looked after the interests of those for whom he felt responsible, and his kindness to those less fortunate than himself – such as the German soldier in St Helena, the by-blows of his uncle and brother, and the émigré Marquis de Chabert. We have seen how he took infinite pains to settle the affairs of two who died while working for him: Gooch the astronomer, and Taylor the computer: and he worked extremely hard to obtain a fair rate of pay for the assistants at the observatory and the computers of the Nautical Almanac in the period of high inflation during the Napoleonic wars.

On the professional side, one of the things that comes clear is the degree of cooperation between Maskelyne and the French astronomers, a collaboration that was greatly appreciated in France. We have already seen how he adopted for the Nautical Almanac the plan which Lacaille had been unable to get adopted in France, and how his lunar distance tables were used directly in the French almanac. In 1792, Lalande said that Maskelyne's ephemerides published in London were by far the best there had ever been. In a long elegy read to the *Institut National* in Paris in January 1813 (we were still at war), Delambre had this to say of Maskelyne's observational work:

In short, it may be said of the four volumes of observations which he has published, that if by any great revolution the works of all other astronomers were lost, and this collection preserved, it would contain sufficient materials to raise again, nearly entire, the edifice of modern astronomy; which cannot be said of any other collection, because to the merit of a degree of correctness seldom equalled, and never surpassed, it unites the advantage of a much longer series of observations; and it must increase in value as it becomes older. . .[3]

The French used these British observations to derive mathematical equations which described the motions of the various bodies in the solar system; Maskelyne in turn used the French equations to improve the predictions in the British Nautical Almanac from 1805 onwards. And all of this despite the wars between Britain and France.

Although he was thus promoting the cause of astronomical science through his routine work at Greenwich – and, most important, making it available to astronomers and mathematicians the world over by ensuring the prompt publication of results – he never lost sight of the principal object of the observatory's existence, namely the improvement of navigation. The stars he chose to observe, his improvements to nautical instruments and development of allied observational techniques, his testing of chronometers, his editing of Mayer's lunar theory and tables – all of these were of the greatest use to the practical navigator. But undoubtedly his greatest achievement was to set in motion the annual publication of Britain's Nautical Almanac, the model for similar ephemerides published all over the world today and the reason why the world's system of time and longitude measurement are today based upon the Greenwich meridian.

APPENDIX A

The Maskelyne pedigree

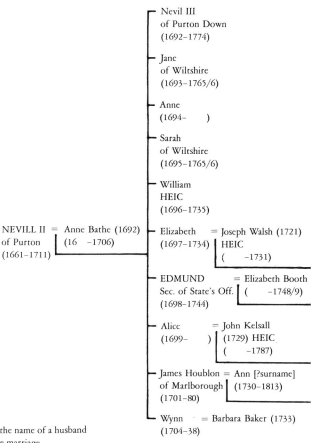

A date in parentheses after the name of a husband or wife gives the date of the marriage.

Principal sources

1. *Burke's Landed Gentry*, 15th ed. (1937), 1547–8
2. The Merriman pedigrees, BL MS Add.39690-2.
3. Nevil Maskelyne's memorandum books, WRO MS 1390/2.
4. Lady Clive's letters, Arnold-Forster MSS.
5. Molly C. Poulter, A descriptive list of the Ormathwaite Collection (1965), IOLR MSS.Eur.D.546.

For earlier and later genealogical information, see 1 and 2 above.

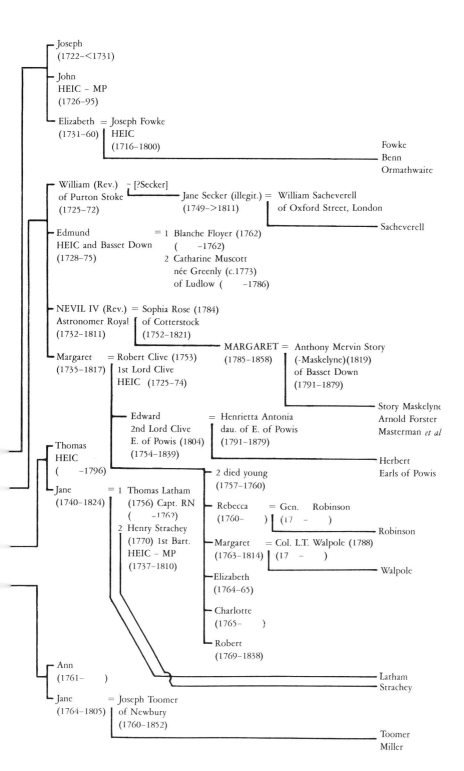

APPENDIX B

Nevil Maskelyne's autobiographical notes

In a folder inscribed by his daughter Margaret:

'Autobiographical notes in Dr. Maskelyne's own hand.'

The text transcribed here is from Maskelyne's original draft in ink (RGO MS.4/320:8), which, from internal evidence, was written in 1800 (Watermark 1798). The original has certain editorial amendments in pencil, probably done after his death by his daughter Margaret.

A verbatim publication has not yet been discovered but it is possible it was prepared for Abraham Rees's *Cyclopaedia*, issued in fortnightly parts from 1802 to 1820: certainly the biography which appeared in Volume 22 (published in 1812) obviously used it as one of the sources. Delambre in his *Eloge* of 1813 thanks Margaret for her help and, from its wording, it seems likely that Margaret sent an edited copy to Paris.

Page numbers in the text in brackets are those of the original manuscript. Maskelyne's insertions and footnotes have been placed where it appears he intended they should go, though this is not always clear.

The transcription
Nevil Maskelyne D.D. & F.R.S
Astronomer Royal

Dr. M. is the last male heir of an antient family long settled at Purton in the County of Wilts, which from the name probably came from Normandy, where there is or was 50 years ago a family of that name Masqueline. He was born in London in the year 1732, received his classical education at Westminster School, & instructions in writing & arithmetic during intervals of school from other masters, which proved of the greatest service to him in his future scientific pursuits. So much time is taken up in attaining the learned languages themselves that it cannot be expected that a great stock of knowledge can be acquired at school for forming the mind; but this he supplied by reading with avidity our best english authors. From occasional discourses in the family, he became eager to see the effects of telescopes on the heavenly bodies, and to know more of the system of the universe. Observing the great eclipse of the sun in 1748 with the late Mr. Ayscough in an unusual manner by means of the sun's image projected through a telescope on a white screen in Camera obscura[1] added fresh spur to his astronomical desires, and from this time he turned himself seriously & closely to the study of the two kindred sciences of Optics & Astronomy, which he has pursued with unwearied diligence ever since. Many mathematicians have become Astronomers from the facility mathematics gave them in the attainment of Astronomy; but here the love of Astronomy was the motive of application to mathematics without which our Astronomer soon found he could not make the progress he wished in his favorite science; in a few months, without any assistance he made himself master of the elements of Geometry & Algebra. With these helps he soon read the principal books in Astronomy & Optics & also in other parts of natural philosophy, Mechanics, Pneumatics & Hydrostatics. [p.2] The considerable progress he had made in these sciences led him naturally to the University of Cambridge, where these studies were then [Deleted in pencil – as they are still,] peculiarly encouraged. In taking his degree he stood the 3d on the list of the first rank of honors,[2] & afterwards was elected fellow of Trinity College. In 1758 he was elected a fellow of the Royal Society.

The expected transit of the planet Venus over the sun in 1761, a phaenomenon which does not happen above once or twice a century, occupied the attention of Astronomers during the preceding year to prepare for the due observation of it, to which they were stimulated by the exhortations of the late Dr. Halley published in the philosophical transactions, who had observed the exit of Mercury from the sun at the Island of St. Helena in 1672 apparently instantaneous & thence concluded that the like phaenomenon of Venus might be observed the same, from which he concluded that if both the entrance & exit of Venus were observed by two observers in proper distant stations on the globe, that great *desideratum* in Astronomy, the sun's parallax, might be determined to the 500th part of the whole. Altho' Dr. Halley was mistaken in some particulars, partly owing to the imperfect Astronomy of the planet at that time, so that the transit of Venus could not be expected to determine the parallax as near[?] as he had supposed of so great advantage, yet still it afforded the best means of determining it that had yet been put into practice. The

Royal Society, therefore, to whom Dr. Halley had in a manner bequeathed this observation, took up his plan with becoming zeal and spirit, & appointed our Astronomer, with an assistant the late Mr. Waddington, to go to the Island of St. Helena to make the observation there as they appointed Mess. Mason & Dixon to go to Bencoolen for the same purpose. The cloudiness with which the Island of St. Helena is so frequently infested, owing to the great height [p.3] and nearness of the hills in so small an island, unfortunately deprived our Astronomer of the important observation of the exit of the planet from the sun's body; which loss was fortunately supplied by the observation being made in good weather by Mess. Mason & Dixon at the Cape of Good Hope, whither they had gone on account of the impossibility of reaching Bencoolen in time, by delay they had experienced thro' the events of war. On this trying occasion he is said to have born his disappointment with so much fortitude as to have said he hoped to meet with better weather to observe the next transit, that which was to happen eight years after, & was indeed better circumstanced for the main purpose of determining the sun's parallax. But the transit of Venus was not the sole purpose of his expedition to this Island. He had proposed to the Royal Society the observation of the distances of the star Sirius from the zenith, which it approaches within half a degree continued for a twelve month, in hopes of finding out an annual parallax; & they furnished him with a zenith sector of ten feet for the purpose; but here he was again disappointed, owing to a faulty method of suspension of the plumbline by a loop from the neck of a pin at the centre, tho' made after the construction of that great Artist Graham, which produced an error many times greater than the parallax could be supposed to consist of, & which it was his fortune or misfortune first to detect; in doing which, however, he may be said to have rendered important service to Astronomy by preventing any other instruments in future from being constructed in so faulty a manner. But he had still in the fertility of his astronomical knowledge & pursuits other resources left to render his voyage useful to Astronomy & to console him for his two disappointments; namely in proving by his astronomical observations of the ☽ dist. à ☉ & stars with the Hadley's quadrant the certainty & utility of this new method of finding the longitude at sea, which he was hereafter to have so great a share in bringing into practice; in his astronomical observations [p.4] for settling the latitude & longitude of the place; in determining the relative force of gravity to that at Greenwich by the quantity of the daily retardation of the astronomical clock; and by his observations of the tides in which he used the assistance of Mr. Mason** but principally by a series of observations of the moon's parallax in right ascension, which, as afterwards he explained it in a memoir presented to the Royal Society, having a relation to the diameter of the parallel of latitude, might serve in conjunction with similar observations made in different latitudes to determine the relative magnitude of several parallels & thence afford a means of determining the true figure of the earth.

**[Mr. Mason], just returned from the Cape of Good hope with Mr. Dixon who had jointly been more fortunate than our Astronomer in being favored with a serener sky & having succeeded in getting a good observation of the going off of Venus from the Sun, which complete the supply[?] the loss of our Astron' obsn. at St. Helena.

Their destination had been for Bencoolen to observe the total duration, but meeting & engaging a french ship of superior force whom however they beat off after the loss of 40 men & another ship of ours coming in sight, they were able to put back to refit & on their setting out again being not likely to arrive in time for the transit at Bencoolen their original destination, thought it best to go to the Cape, a place equally well situate for finding the sun's parallax from the exit of Venus from the sun's disc, as St. Helena. As he availed himself of Mr. Mason's assistance to make a capital experiment of six weeks of the height & times of the tides at St. Helena, so his zeal caused him to invite Mr. Dixon to return to the Cape of Good Hope, taking the Astronomical clock with him which himself had used in determining the difference of gravity between Greenwich & St. Helena, & other instruments, in order to make the like experiment of gravity at the Cape, & further obs' of $\jupiter$ sat. to settle the long. of the place, [p.5] which observations, however, we believe have never been published.

While he was passing a day making some observations at the opposite part of the Island to St. James's Fort, he was accosted by a soldier of the East India Company doing duty at the Fort there, who appeared by his behavior & address to be above the common sort, who informed him he was a German & being reduced to poverty in London had been induced under some deception as he thought to engage to go out in the company service as a soldier in St. Helena, & that he wished much to be released from the service in order that he might return to England & thence to his own country, where he could get into a better situation in life. Our Astronomer immediately made use of his interest with the then worthy

Governor of the Island, Mr. Hutcheson, to procure the man's discharge, and to effect it paid down the levy money requisite to procure another soldier in his stead.

During his stay at St. Helena, he met with the greatest civilities, from the courtesy & friendliness of the Governor (the reverse of what his predecessor Dr. Halley at the same place had met with in 1676 from the vexatious tyranny of the Governor of that time, whose name is best buried in oblivion[3]) and also every assistance towards building his Observatory & other accommodations, agreable to the recommendations sent out with him from the East India Company, who were pleased liberally to patronise an expedition undertaken by the Royal Society for the honor of the nation and the benefit of Astronomy & Navigation, in which too the Company were so highly interested.

[p.6] During the voyage to St. Helena, both out & home, he made continual observations for the longitude by observations of the moon with Hadley's quadrant then first rendered practicable by the publication of the first edition of Mayer's lunar Tables in the Gottingen Acts, & by the improvements of instruments, and instructed the officers of the ship both in the observations & calculations, who became the first in that service who practiced the same, which since has become a general part of the business of the company's ships, as well as practiced on board his Majesty's ships & other services.

[p.5] Soon after his return to England in the spring of 1762, being introduced to the Court of Directors, he acknowledged his obligations [p.6] to them for the civilities he had received in consequence of their directions, in a handsome speech, and at the same time congratulated them on the prospect of a great improvement in the navigation of their ships that now offered itself to use by the lunar method of finding the longitude at sea, which he had proved the practicability of in his late voyage on board their ships. The chairman did him the honor to give a polite answer to his speech, & to express the satisfaction of the Court with the contents of it, & with his exertions to promote the improvement of navigation in so essential a point.

At this time Mayer's Tables were only to be found in the Gottingen Acts, the nautical almanac did not exist, nor had any treatise yet appeared containing a description of a plain & easy method of computing the longitude from observations of the moon's distances from the sun & fixed stars taken at sea, level to the capacity and use of sailors. Our Astronomer in the year 1763 supplied these defects, by the publication of the British [p.7] Mariners guide, containing at the same time an english Edition of

Mayer's Tables & directions for making the necessary observations and calculations. With these helps, several Commanders & mates of ships in the service of the East India Company, the very next year, corrected their reckonings in their voyages to the East Indies & China within a degree of the truth; which they testified on examination before the Board of Longitude in the beginning of the year following.

The indefatigable and accurate Dr. Bradley, Astronomer Royal, had departed this life in August 1762, at the age of 69, a life probably much shortened by his laborious calculations to verify Mayer's Tables by a comparison of them with 1200 of his observations, during 5 years from the end of 1755 to the end of 1760, an event much to be deplored, and which must still be considered as a great loss to Astronomy, as the observations made at the Royal Observatory must have been rendered more valuable by his longer cooperation in & superintendance of them, & he might have taken some measures for communicating them to the learned world when rendered more complete in the manner he wished both as to the observations & calculations. Mr. Professor Bliss was immediately, thro' the recommendation of the Earl of Macclesfield, President of the Royal Society, appointed to succeed him, who tho' incapable from his age and infirmities from taking any active part in making the observations directed his Assistant, who had been the same to Dr. Bradley, to keep on a regular series of them, not deficient in accuracy tho' less complete than formerly as to the number of observations. In the mean time, Dr. Bradley's Executors claimed the books of his observations as the private property of his representatives. Our Astronomer alarmed with the apprehension of the loss which himself with the whole astronomical [p.8] world were likely to suffer from the effects of this claim, addressed the Royal Society on 9th June 1763 recommending it to them to pursue proper measures for recovering the observations as public property from the hands of the Executors for the use of the public; in which he was seconded by the Earl of Morton, who the next year was elected President of the Royal Society on the decease of the Earl of Macclesfield. The Society resolved to take measures for recovering the Observations for the use of the public; & to print them when recovered at the expence of the Society; & to call upon the Astronomer Royal every year, in future, for his observations, agreable to a right given them to that by a letter of Queen Ann, & to print them yearly in the philosophical transactions. These would have been great & glorious advantages to Astronomy, if they could have been carried into effect. But the Executors persisted in detaining the observations, & the right given to the Society by Queen Ann's letter was denied having any validity after her decease. In fine, three years afterwards the Earl of Morton, the new President of the Royal Society, brought the matter before the Board of Longitude (of which he & our Astronomer were members) who joined in recommending it to the Secretary of State & thro' him to his Majesty, to demand the Observations as the property of the crown for the use of the public (at whose expence they were made) & in case of refusal to institute a suit in the Exchequer to compel the delivery of them. The opinions of the Crown lawyers being in favor of the right of the Crown, a suit was commenced accordingly for the recovery of the observations, but the Earl of Morton dying soon after[4], it was carried on tardily, & at last in 1775 the Executors thought proper to make a present of them to Lord North at once Chancellor of the Exchequer & Chancellor of the University of Oxford, who immediately made a present of them to the University of Oxford on the condition of their printing & publishing them. They were immediately put into the hands of the Professor of [p.9] Astronomy at Oxford[5] for that purpose; however, it was not till two years ago [1798] that the first Volume of it in Folio was published, owing as it is said to the ill health of the Professor; who we hear has since given up the continuance of the work. It was then put in the hands of the Savilian Professor of Geometry[6], who made some progress in the printing, but has since thought it proper to give it up likewise. The learned world still waits anxiously for the publication of the second Volume, which would complete the more valuable part of Dr. Bradley's observations; we are sorry to add that rumors are about that some demur or stoppage at present retards the progress of this valuable work. We flatter ourselves that our readers will not be displeased at our giving the continuance of the history of these observations, so interesting to Astronomy, the first steps for the recovery & publication of which were made by our Astronomer.[7]

History of Mr. Harrison's time-keeper, & of Mayer's Tables.

The Board of Longitude being about to send out Mr. Harrison's celebrated time-keeper, with the care of his son, on a second voyage to the West Indies, for a further trial of it (as some doubts had occurred with respect to the success of the trial in the voyage to Jamaica) our Astronomer was solicited to undertake the voyage, and the island of Barbados was fixt on for the place, & he was instructed to make trial of Mr. Irwin's marine chair on the voyage, and to make observations of equal altitudes of the sun on the arrival of the watch, for comparing it with the meridian of the place, and other observations, in conjunction with Mr. Charles Green late Assistant at the Royal Observatory, for ascertaining its Longitude. He found Mr. Irwin's marine chair was too much disturbed by the motions of the ship to allow the management of a telescope sitting in it to observe the eclipses of Jupiter's Satellites. Harrison's watch was [p.10] found to give the Longitude of the Island with great exactness.

[p.9] Besides performing these services, he in the voyage out & back repeated the observations of the distances of the moon from the sun and fixed stars whereby he further confirmed the certainty and utility of that method; and extending his stay at the Island to near a twelve month, he made necessary [p.10] observations of the eclipses of Jupiter's satellites, occultations of fixt stars by the moon, and of the moon's diurnal parallax in right ascension, which latter he had commenced at St. Helena, but had now an opportunity in a better & more serene climate to multiply and render more complete than the cloudy sky of St. Helena allowed him.

On his return to England in the Autumn of 1764, he found that Mr. Bliss the Astronomer Royal was lately deceased, after enjoying the office only two years; and he had been already strongly recommended to his Majesty's ministers by his friends, & among them the Earl of Morton President of the Royal Society; to these were afterwards added the principal members of the Royal Society and the Professors of Mathematics and Astronomy of the University of Cambridge, to which he belonged. On the other hand, he was opposed by a gentleman of the University of Oxford who was supported by the powerful weight of that University. The superior merits of our Astronomer[8] turned the balance of Royal patronage in his favor, and he obtained the nomination to this honourable office.

The very morning he received the appointment had been nominated for a meeting of the Board of Longitude, to take into consideration the late trial of Mr. Harrison's time-keeper, and to decide upon his claim to the premium held out by Act of Parliament. The President of the Royal Society proposed that £10000 should be given to Mr. Harrison immediately on discovering the construction of his time-keeper, with this further stipulation in his favor, that as soon as two more should have been constructed by other artists to prove the practicability of the invention & they should have been tried at the Royal Observatory & on voyages at sea to prove this method generally practicable & useful, he should be intitled to the other £10000. This was unanimously agreed to. The Astronomer Royal then came forward [&] gave his evidence of the utility & certainty of the lunar method of finding the Longitude, from [p.11] his own experience, & that of several Captains & mates of ships in the service of the East India company, some of whom appeared in person & were examined, & others sent their testimony in writing; and proposed that the Board should order a nautical and astronomical almanac to be constructed & published to facilitate the calculations of the sailors in finding the Longitude at sea by the lunar method. In conclusion the board approved of this no less than of the proposals of the Presidt. of the Royal Society, & agreed to apply to Parliament, for authority to pay to Mr. Harrison the sums before mentioned on the conditions before recited, & to grant money to purchase Mayer's Tables & to defray the expence of computing & publishing a nautical almanac; a work which every lover of Astronomy, Navigation & Commerce would readily say, with Emphasis, Esto Perpetua! This was the beginning of that highly useful publication, the nautical almanac; the first of which was that of 1767. The Astronomer Royal not contented with seeing the nautical almanac established with the sanction of Parliament, under the care of the Board of Longitude, voluntarily undertook to superintend the computers of it & give them such instructions as they might stand in need of; and in short to be the editor of it; which he has continued ever since, the last published being that of 1804, it being thought proper to publish it a few years in advance.

On the petition of the Royal Society, his Majesty was pleased to renew the letter formerly given by Queen Anne, appointing them Visitors of the Royal Observatory, with the right to demand a copy of the observations made there every year; & His Majesty ordered them to be printed & published at the public expence under the direction of the Society. Those made by the present Astronomer have accordingly been all printed and published to the end of the year 1798, and are to be continued annually.

[p.12] It may not perhaps be amiss, here, to say something of the accuracy of the instruments at the Royal Observatory & of the observations made by them. The instruments which had been made use of by Mr. Flamstead, not having been paid for by Government, were taken away by his Executors. Dr. Halley obtained a grant from Government of an excellent mural quadrant of 8 feet radius made by Sisson under the direction of the celebrated Graham, and divided by Graham, or Sisson, and two or three pendulum clocks of Graham's construction; who, at the same time that he was an excellent clockmaker, possessed that uncommon genius & taste for mechanics that he invented & perfected some of the principal astronomical instruments. He had before in 1725 invented & made a zenith sector of 25 feet radius for Mr. Molyneux, & then one of 12½ feet radius for Dr. Bradley, which enabled the latter to observe the phaenomenon and laws of the aberration of light, and thence to discover its cause, & afterwards to discover the nutation of the earth's axis. When Dr. Bradley succeeded Dr. Halley at the Royal Observatory in 1742, he improved the mural quadrant by the addition of a divided micrometer circle shewing seconds, to the screw which moved the telescope in the time of observation to bring the star upon the wire; & he constantly made use of a plumbline to set it to the perpendicular, which had been neglected by Dr. Halley; & obtained compound or gridiron pendulums to be added to the clocks, an ingenious invention of Harrison. It is probable that magnifying glasses fitted in tubes were then first applied by him to observe the bisection of a fine point upon the arch by the plumbline for rectifying the quadrant to the perpendicular, & to observe the coincidence of the line of the Vernier with the line of division on the limb, in reading off an observation. Dr. Bradley also possessed a transit instrument having a telescope of six feet long and an axis of [3½] feet, left by Dr. Halley, which had the [p.13] great defect of not having the telescope in the middle of the axis, probably so made in order to adapt it to the narrowness of the place where it was thought proper to fix it. He had also a telescope of 15 feet with a micrometer of his own.

These instruments he made the best use of till the years 1749 & 1750, when he got a grant of a new sett of instruments from Government, viz. a brass mural quadrant of 8 feet radius, & a transit instrument of the same length, and a moveable quadrant of 40 inches radius, made by Bird; a Newtonian reflecting telescope of six feet by Short; and an astronomical clock by Graham. He also got two instruments of his own to be made a part of the apparatus of the Royal Observatory, viz. the zenith sector of 12½ feet radius before mentioned, and an equatoreal sector with a telescope of 30 inches for observing comets, also made by Mr. Graham. The Astronomer Royal has made several improvements to these instruments at various times. In 17[75], gold points were fixed upon the central plate & arch of both quadrants for their rectification perpendicular by the plumbline; achromatic object glasses and moveable eye-glasses were applied to the transit instrument and brass mural quadrant in 1772, and the magnifying power of the 1st increased from 50 to 82 & of the second from 50 to 63, and an achromatic object glass to the old iron quadrant in 17[89]; in 1768 the suspension of the plumbline of the zenith sector was altered from a notch placed at the centre to one placed above the center of the instrument & moveable from right to left by a screw to adjust the plumbline to a fine point in gold placed at the centre of the instrument. In 1785 the brass arch of the zenith sector was exchanged for a steel one, with the divisions on gold points, to render the expansion of the arch by heat & contraction by cold conformable to those of the tube of the telescope which forms the radius of the instrument & is composed of iron plates turned over. Wedges of green & red glasses with their refracting angles placed to contrary ways have been introduced to be interposed between the eye & the telescopes to darken the sun in making observations of him, which are [p.14] far preferable to smoaked glasses, both in distinctness and in permanency.

APPENDIX C

Maskelyne's expense accounts, St Helena, 1761–3

Both in his own hand.

Submitted before departure, 1761
The Council of the Royal Society Dr.
to the Revd. Nevil Maskelyne –   L   s   d

To Money advanc'd to Mr. Sisson on account of the 10 foot Sector	79.18. 2
3 Gallons of Oil to burn in the Use of the Sector 11s 0d. Bottle & Basket for Do. 2s 6d	0.13. 2
Pd. to Mr. Waddington for Porter's carrying the Clock which Mr. Shelton made from the Royal Observatory at Greenwich to the Waterside 2s. Wateridge from thence to London 4s. Porters from Waterside in London 2s 6d. Beer 1s	0. 9. 6
Box maker opening the Chests at the East India Wharf, & nailing them down again	0. 7. 6
Pd. Mr. Fleet Clerk at the Pay Office at the India House, the Customary fees for upwards of 30 cases at 1s per case	1.11. 6
Pd. Mr. Waddington. Porters for Loading the 3 cases belonging to the Sector into the Cart to go from Mr. Ellicott's to Botolph Wharf 2s 9d. Carriage on a Horse for 4 Cases containing the Transit Instrument and Telescopes 3s 0d. Porterage of Clock from the Minories (where it had been left when it was brought from Greenwich) to Botolph Wharf 2s 6d. Cart for carrying 3 Cases belonging to the Sector 5.6	0.13. 0

226 *Appendix C*

Cart for carrying my Boxes from Grey's Inn to Botolph Wharf 5s. Carter 1s. Porter that went with the Cart 2s 6d. Beer 1s	0. 9. 6
	84. 2. 4
Recd. of Mr. Davall in order to advance to Mr. Sisson for making the Sector	50. 0. 0
Ballance due to me	34. 2. 4

(Ordered to be paid at RS Council December 23 1760)

(RS Misc.MSS 10/148)

Submitted after return

The Council of the Royal Society Dr. to the Revd. Nevil Maskelyne F.R.S.

	L s d
To a bill of charges paid at the India House before my setting out in Janry 1761 including 1£ 2s paid Mr. Sisson for work done to the dipping needle, & a few other articles	8. 4. 5
Wine Merchant's Bill	56.10. 0
A Hogshead of Porter and Cask	3. 4. 0
5¼ Gallons of Arrack	1. 2. 6
3 Gallons of Lemon Juice	0.10. 0
Candles I took out with me. 10Lb of Wax & 10Lb of Moulds	1.15. 2
½ a hundred weight of soap	1. 7. 6
Sugar 12 loaves I took out with me	5. 9. 6
Tea 12 Pounds I bought at St. Helena	3.12. 0
Expence of my journey and Mr. Waddington's to Portsmouth in January 1761	5.11. 3
Expences at Portsmouth	0.18. 3
Paper, pens, writing books and ink	1. 0. 8
Carriage of Buroe and other things from London to Portsmouth	1. 6. 6
Porterage at Portsmouth	0. 7. 0
Porters bringing instruments &c on shore at St. Helena	1. 7. 6

Maskelyne's expense accounts, St Helena, 1761–3 227

Expence of servant whom I took to wait on me at St. Helena	10.10. 7
Barber at St. Helena	1. 1. 3
Washing at St. Helena	3.19. 6
Workmen	3.16. 0
Customary gratuities to servants on board of ship and in families where I lived and visited	8.12. 0
Journey from Plymouth to London	7.14. 4
Carriage of my box from Plymouth to London	1. 3. 6
	129. 3. 5
Brought over	129. 3. 5
Mr. Dollond's bill for a 10 foot object glass sent to me out in St. Helena	1. 1. 0
Mr. Sisson's bill for silver wire sent me out to St. Helena and for fitting the said object glass in sliding cells for the Sector	1.15. 0
Lodging and boarding at St. Helena from April 6 1761 to Feb 19 1762 being 319 days at 6s per day, as per agreement with the council	95.14. 0
Sum which the council agreed to give me for my services	150. 0. 0
Made Mr. Dixon a recompence for returning to the Cape of good hope to set up the Society's clock there (which I carried out with me) in order to determine the difference of gravity between that place & St. Helena	10. 0. 0
Made Mr. Mason a like present for his assistance in observing the tides at St. Helena	10. 0. 0
Sum Total. The Council of the Royal Society Dr. to the Revd. Nevil Maskelyne	397. 3. 5
Per Contra the Revd. Nevil Maskelyne Dr. to the Council of the Royal Society for 160L received by him on account for Liquors, and 80L drawn by him upon the Council from St. Helena, and 20L from Plymouth – Deduct	260. 0. 0

    Remains ballance due from the Council of the Royal
       Society to the Revd. Nevil Maskelyne   137.13. 5
    [In Davall's hand]
    Received of Mr. Davall this 21st day of July, 1762
       The sum of one hundred and thirty seven pounds thirteen
       sh & 5d being the balance of the above account
       I say received the same by me
       [signed] Nevil Maskelyne

                                      (RS, James West Accounts papers)

The wine merchant's bill for £56.10.1 referred to above has been preserved. from Wiliam Wright, dated December 17 1760, it included 16 gallons of Madera, £8.0.0; 16 gallons 2 quarts of Mountain, £5.10.0; 65 gallons of Lisbon and port, £20.11.8; 6 dozen Claret, £14.11.9; and 10 gallons of Rum, £5.0.0. The remaining £10.11.2 of the gross sum was accounted for by casks, bottles, porterage, packing, etc., and then a credit of £7.14.6 was given as Draw Back of Duty on wine and bottles.

                                                     (RS Misc.MS 10/151a)

APPENDIX D

The 1765 inventory

This inventory, taken at Maskelyne's first observatory Visitation on 16 March 1765 (RGO MS 6/21, 78–80), lists the instruments and apparatus for which he became responsible. Editor's remarks are in *italic*.

1 Transit Room *[see Fig.7.3]*
1. an Eight feet transit Instrument with its levels. *[Bradley's transit, by John Bird, 1750.]*
2. A Month Clock with a Compound Pendulum. *[Bradley's transit clock by John Shelton, 1750; known as 'Graham 3'.]*
3. A Parallactic Sector of 12½ feet radius: this sector can be moved into the Quadrant room. *[Bradley's zenith sector, by Graham, 1727, with which he made the discoveries of aberration and nutation: used now only to check the accuracy of the mural quadrants.]*
4. a Pair of Globes 17 inches Diameter.

2 Quadrant Room
1. A Brass Quadrant of 8 feet radius fixed to the west side of the Pier. *[Bradley's mural quadrant, by Bird, 1750; facing south.]*
2. an Iron Quadrant of 8 feet radius fixed to the west side of the Pier, having a brass arch. *(Halley's mural quadrant, by George Graham, 1725; re-divided by Bird, 1750; facing north.]*
3. A Month Clock with a compound pendulum. *[Halley's 'Graham 2', 1725; gridiron pendulum by Graham, 1743.]*
4. Two Barometers & a Mercurial Thermometer.
5. A Board with two pieces of brass screwed to it, to try the quantity of the arcs of the two Eight feet Quadrants.
6. a Brass arch which was made to try the line of Collimation of the Telescope of the Iron Quadrant.

3 In the middle Room

Two Alarums: one in the study; the other in the room above.

4 In the Great Room

1. A Moveable quadrant of 40 Inches radius with all its Apparatus. N.B. there is a room lately compleated, on the South side of the Observatory, for the reception of this Quadrant. *[This was a scaled-down version of the mural quadrants, rotating on a vertical shaft so that zenith distances could be measured on any bearing; by Bird, 1754.]*
2. A 6 feet Newtonian Telescope. *[By James Short, 1756; for Jupiter's satellites, etc.]*
3. A decimal *[sic for equatorial]* Sector with the Stand & apparatus for adjusting it, & one Spirit level. N.B. the other level is wanting. *[Bradley's 30in. equatorial sector by Graham, c.1735; for measuring the positions of comets by comparison with nearby fixed stars.]*
4. A Month Clock with a Compound pendulum. *[Halley's 'Graham 1', 1725; gridiron pendulum by Graham, 1743.]*
5. A Week Clock, with a simple pendulum. *[Halley's transit clock, by Graham, 1721.]*
6. A 15 feet Telescope with its Micrometer, object, & eye glass of 7 feet focus. N.B. another object glass of 10 feet focus mentioned in the former wanting & said to be broken. *[Bradley's refractor tube, used with 7, 10 or 15 feet objectives, brought to Greenwich in 1742.]*
7. A dipping needle with its apparatus, three needles.
8. An Horizontal needle with its apparatus, three needles, three variation Compass boxes. *[There is no record of Maskelyne having made any magnetic observations.]*
9. A four feet common Telescope.
10. Two old Micrometers. *[Halley's]*

5 West End of the House

A five feet transit Instrument & its levels. *[Halley's transit instrument, with telescope possibly by Robert Hooke, before 1721.]*

6 In the Woodhouse

1. A 20 feet refracting Telescope tube in one piece. *[Probably Halley's.]*

2. Two old wooden tubes.
3. A Wooden apparatus in the form of a long Box for finding the line of Collimation of the telescopes of the 8 feet quadrants.
4. A 24 feet wooden tube. *[Halley's.]*

APPENDIX E

Schiehallion instruments and equipment

Astronomical instruments
Zenith sector, 10ft radius, 17° arc graduated to approximately 1 arc-minute, achromatic object glass. Made by Jeremiah Sisson in 1760 for Royal Society. With Maskelyne in St Helena 1761–2. Plumb-line suspension modified and stand provided to permit reversal east to west 1771–2. Present whereabouts unknown.
Tent for sector, parallelopiped, 15.5ft square and 17ft high, wood joists covered with painted canvas.
Clock, astronomical, made by John Shelton 1760 for Royal Society for 30 guineas. With Maskelyne St. Helena 1761, Barbados 1763, Mason & Dixon 1765–6. Present whereabouts: Royal Museum of Scotland.
Astronomical quadrant, 1ft radius, made by John Bird for 1761 or 1769 transit of Venus.
Transit instrument, made by John Bird for 1761 or 1769 transit of Venus.

Surveying instruments
Theodolite, 9in diameter, capable of observing angles to ±1 arc-minute, by Jesse Ramsden.
Theodolite, lent by James Stuart Mackenzie.
Two *portable barometers*, Deluc-type, with staff and bag, by Edward Nairne, 1773.
Gunter's chain.
Painted tape, 3 poles long, marked with feet and inches.
Two *fir poles* of 20ft each, and four wooden stands. Provided in Scotland.
Brass standard, 5ft long, made by John Bird 1766 for Mason & Dixon
Two *thermometers*

Camping equipment
1 tent bedstead & covering
1 ship stove compleat
1 tea kettle; 1 boiling pot; 1 frying pan; 1 sauce pan
1 gridiron; 3 pewter dishes; 6 d.plates
1 pr. bellows; 1 trivet; 1 coal tub; 1 pail
2 deal tables; 3 flap-to stools
6 knives and forks; 6 pewter spoons
1 hatchet; 1 saw; 1 hammer; 1 chissel; 1 handvise
2 gimlets; 1 pr. pincers; 1 pr. pliers; nails and screws
1 pickax; 1 spade; 1 shovel
60 yard of ¾ inch rope; 60 yard of ½ inch rope
24 screw hooks
20 iron spikes, ½ in. square, eyes at upper ends, lower ends case-hardened and 16 inches long
30 yards of tarpaulin
12 garden matts
Sources: Royal Society Council minutes 1773–4.
Phil. Trans. R. Soc. 65, (1775), 500–1.

Maskelyne's Expenses in his Account Book for 1774 (A/C 1/40)

Total of expences on expedition to Scotland to measure the attraction of Schihallion. Accounted for thus nearly	212. 0.11
Setting up & removing the Observatory & instruments & Carpenter's work done about them, building Bothay on N. side of hill, & paying Wm. Menzies for surveying	89. 7.10
Pd. price of a Bothay that was burnt & things contained in it	3. 4. 1½
Provisions on Schihallion & bringing them up the hill	30. 5. 3
Provost John Stewarts bills for sundry necessaries sent from Perth, including wine & other liquors	29. 1. 6
Expence of journeys & Horses	57. 6. 8
	209. 5. 4½
Deduct produce of sale of several articles	5. 8. 0
	203. 7. 4½

234 *Appendix E*

Have recd. of Royal Society by bills of Samuel Wegg Esqr. 146L.3S.8D they agree to pay me the remainder to 212.0.11 viz.65L.17S.3D to make up the full amount of my expences I received this sum (65.17.3) of the Royal Society Janry. 12

[The whole experiment cost the Royal Society (out of the King's grant) £597.16, including £150 to Hutton for his analysis.]

APPENDIX F

Memorial tablets in the Church of St Mary, Purton, Wiltshire

Tablets in the Chapel of St Nicholas, south transept

To Nevill Maskelyne, 1611–1679
(Nevill I – the astronomer's great-great-grandfather)
Depositum
NEVILL MASKELYNE
Arm^{ri} Obijt 30ma Aug^{ti} A^o MDCLXXIX
Eheu: Nec Pietas moram Rugis Afferet
(Alas: even piety cannot delay wrinkles)
A shortened version of Horace's Ode to Death (Odes, 2, 14)

To Nevil Maskelyne, D.D., F.R.S., 1732–1811
(Nevil IV, the Astronomer Royal)

(Latin original)	(Translation by Lesley Murdin)
In hujus Ædis Cæmeterio,	In the cemetery of this church
Depositæ sunt Relliquiæ	Are deposited the remains of
NEVIL MASKELYNE S.T.P. R.S.S.	NEVIL MASKELYNE D.D. F.R.S.
Necnon Astronomi Regii,	Astronomer Royal,
VIRI,	And whether you look at
Seu Morum Simplicitatem,	The simplicity of his way of life
Seu Animi Benevolentiam,	Or the kindness of his heart
Seu Vitæ doctissimæ Utilitatem.	Or the usefulness of his life of learning,
Spectes,	This man is worthy
Insignis, Eximii, publice deflendi	Of being publicly mourned
Qui Naturæ Leges exprimendo,	Since he worshipped the Great Creator
Auctorem Maximum potissimé coluit:	By formulating laws of Nature:
Sine fuco Pius,	Virtuous without pretence,
Pietatem officio exhibuit	He demonstrated goodness in his work,
Deniqui, non Sibi, sed Christo fidens,	Ultimately trusting not in himself,
In gremium Patris, Sempiterni,	But in Christ, to the
Vitam bene actam	Bosom of the Eternal Father
Futuri certâ Spe præmii,	He rendered a life well lived
Reddidit:	In the certain hope of the reward to come.
Quinto Id: Feb. Anno Domini 1811.	This on the fifth of February AD 1811.

On the tomb in the churchyard on the south side of the church

South face of tomb
HERE are DEPOSITED
THE REMAINS OF NEVIL MASKELYNE
DD. FRS.
ASTRONOMER ROYAL, FORTY-SIX YEARS.
He died on the 9th day of February 1811,
at the ROYAL OBSERVATORY in
GREENWICH PARK,
in the 79th Year of his Age.

East face of tomb
IN THE SAME VAULT
ARE DEPOSITED THE REMAINS
OF SOPHIA MASKELYNE,
WIDOW OF
NEVIL MASKELYNE.DD
ASTRONOMER ROYAL.
SHE DIED FEB^Y 15TH 1821,
IN HER 69TH YEAR.

West face of tomb
HERE ALSO ARE DEPOSITED
THE REMAINS OF
MARGARET STORY MASKELYNE
DAUGHTER OF NEVIL AND
SOPHIA MASKELYNE,
AND WIFE OF
ANTHONY MERVYN STORY
MASKELYNE ESQ. OF
BASSETT DOWN HOUSE;
WHO DIED FEB. 15. 1858.
AGED 72.

Glossary

This glossary only contains definitions of the technical terms used in this book, and the definitions have been written strictly within that context. It does not set out to be a comprehensive glossary.

Aberration of light. The apparent displacement of a star from its true position on the celestial sphere due to the finite velocity of light from the star, combined with the velocity of the Earth in its orbit. A similar phenomenon occurs in a moving car when vertical raindrops appear to be approaching from ahead.

Achromatic telescope. A *refracting telescope* fitted with an achromatic *object glass*, which is a compound lens made of two or more components (of different kinds of glass) to overcome the effects of *chromatic aberration*. Peter Dollond designed a very successful example, first tried at Greenwich, of a *triple-achromatic telescope*, whose object glass comprises three elements.

Altitude, in astronomy. The angular distance of a body above the horizon. Altitude is 90° minus *zenith distance*.

Astronomical unit. A unit of length equal to the mean distance from the Earth to the Sun, approximately 149 600 000 km.

Bode's Law, see Titius–Bode Law.

Cassegrain. A form of *reflecting telescope* where the image is viewed from behind the main mirror through a central aperture, as with a *Gregorian* (though they are optically different).

Chromatic aberration. With a simple lens, blue light comes to a focus closer to the lens than red light, with the result that an image of white light, which is a combination of all colours, will be blurred by coloured fringes of out-of-focus light. This defect is overcome in *achromatic telescopes* with compound lenses.

Chronometer, marine. A timekeeper for navigational use, designed to keep accurate time for long periods at sea, regardless of changes of temperature or motion of the ship.

238   *Glossary*

Deck watch. In a ship, it is very bad practice to move a *chronometer* from its normal stowage down below. For timing observations on deck, therefore, a deck watch is used, which can then be taken below for comparison with the chronometer.

Declination, in astronomy. Angular distance of a body from the celestial equator measured positive north and negative south. Declination is usually given in combination with *right ascension* or hour angle to define a position on the celestial sphere. Declination is 90° minus north *polar distance*.

Declination, magnetic. *See Variation, magnetic.*

Dip, magnetic. In nautical parlance, the vertical angle below horizontal of a freely suspended magnetic needle at any place. To physicists, it is known as *magnetic inclination.*

Dip circle or *Dipping needle.* Instrument for measuring the quantity of *dip.*

Diurnal movement. The apparent movement of a heavenly body across the sky during the course of the day.

Eclipses of Jupiter's satellites. The predicted Greenwich times that Jupiter's Galilean satellites pass into the planet's shadow are given in the Nautical Almanac; if these are compared with times actually observed by telescope, longitude can be found.

Ephemeris, plural *Ephemerides.* A table of predicted positions of a celestial body at a series of times, generally published in an almanac.

Equal altitude instrument. An instrument for taking equal altitude observations for time determination, by timing the moment a body is at the same *altitude* some hours before and some hours after it crosses the meridian; half way between the two gives the moment of meridian *transit* with precision, with the sun, it gives noon. An *astronomical quadrant* was generally used for that purpose in Maskelyne's time.

Equation of time. The difference between apparent solar time (sundial time) and mean solar time (clock time).

Equatorial mounting. A method of mounting a telescope with one axis of rotation (the polar axis) parallel to the Earth's axis. Once the telescope is pointing at a star, it can be kept in the field of view by rotating the telescope around the polar axis only.

Figure of the Earth. Broadly speaking, this is the amount by which the shape of the Earth departs from the sphere, being flattened at the poles – an oblate spheroid. Maskelyne made gravity measurements by comparing the rate of the clock in different latitudes to help determine the Figure of the Earth.

First point of Aries. The zero for measurements of *right ascension*, the point of intersection of the celestial equator and the ecliptic, at or near the point where the Sun crosses the equator from south to north on about 21 March. It is also known as the vernal equinox.

Great circle. A circle on the surface of a sphere whose plane passes through the centre of that sphere. The shortest distance between two points on the surface of a sphere is along the great circle between them.

Greenwich apparent time. Apparent solar time (sundial time) on the Greenwich meridian.

Greenwich mean time (GMT). Mean solar time (clock time) on the Greenwich meridian.

Gregorian. A form of *reflecting telescope*, where the image is viewed from behind the main mirror through a central aperture, as with a *Cassegrain* (though they are optically different).

Inclination, magnetic, See *Dip*.

Internal contact. During the transit of a planet across the disk of the Sun, internal contact occurs at the moment when the *limbs* of the planet and Sun are touching each other, with the whole planet silhouetted against the Sun's disk.

Limb, in astronomy. The apparent edge of any celestial body having a detectable disk.

Meridian. To an observer at a particular position on Earth, the meridian is his north–south line.

Meridian altitude. An *altitude* observed at the moment a body is on the meridian. This is the simplest way for a navigator to find his latitude, knowing the *declination* of the body.

Micrometer, astronomical. An instrument attached to a telescope for measuring angular distances between two objects, both in the field of view.

Moon's age. The Moon's phase, expressed in days after New Moon.

Newtonian. A form of *reflecting telescope* where the eyepiece is on the side of the tube.

Nutation of the Earth's axis. The regular 'nodding' of the direction of the Earth's axis in space, caused mainly by the Moon's gravitational attraction.

Object glass. A lens, forming the *objective* of a refracting telescope.

Objective, telescopic. The lens (for a refractor) or mirror (for a reflector) which gathers light from a distant object and forms an image at the focal plane of the objective.

Occultation. The obscuration of one celestial body by another of greater apparent diameter, especially the passage of the Moon in front of a star or planet. The opposite to a *transit* (2). Such observations can yield the observer's longitude.

Parallax. The difference in apparent direction of an object as seen from two different locations; conversely, the angle at the object that is subtended by the line joining two designated points. In nautical astronomy, parallax has to be allowed for, particularly with the Moon, because the observer is at the Earth's surface whereas the ephemerides are calculated as if he were at Earth's centre (*geocentric parallax*). Observing the transit of Venus was an effort to measure *solar parallax* to find the sun's distance from Earth. The radius of the Earth's orbit (or *astronomical unit*) is used as a baseline for measuring *stellar parallax* (observations of the same star six months apart should show a difference of direction in space) to find stellar distances, though, because of the enormous distances involved, this can only be done for the nearest stars.

Polar distance. Angular distance on the celestial sphere between the body and the north or south celestial pole. North polar distance is 90° minus *declination*.

Quadrant. A quarter of a circle (90°), hence an instrument for measuring angles, generally with an arc of 90°.

Quadrant, astronomical. In Maskelyne's time, generally refers to a small portable quadrant taken on expeditions, usually set up vertically to measure *altitude* or *zenith distance*, the vertical being defined by plumbline.

Quadrant, movable. An observatory instrument in the same form as the above, set up vertically and able to rotate about a vertical axis, so that it can point to any *azimuth*.

Quadrant, mural. A quadrant mounted vertically on a wall, generally in the meridian, not free to move in azimuth. One of the principal observatory instruments in Maskelyne's time, for measuring *zenith distance*, and hence obtain *declination* or *polar distance*.

Quadrant, reflecting. A nautical instrument for measuring angular distances up to 90° by double reflection. Known as *Hadley's quadrant*, it is later called an *octant*. From it was developed the nautical *sextant*, measuring up to 120°.

Refraction, atmospheric. The change in the direction of travel (bending) of a light ray as it passes obliquely through the atmosphere. As a result, the observed *altitude* of a body is greater than its true altitude, the amount depending upon the altitude.

Right ascension. Angular distance on the celestial sphere measured eastwards along the equator from the *First Point of Aries*. It is usually expressed in units of time and given in combination with *declination* to define a position on the celestial sphere.

Running fix. In navigation, a fix is a position obtained by observations of two or more objects (terrestrial or celestial), generally made more or less simultaneously. Where there is a significant interval between observations, it is necessary to allow for the ship's movement ('run') during that interval, and the result is known as a running fix.

Sector, equatorial. An instrument on an *equatorial mounting* for measuring angular distances at any inclination, generally used for finding the position of a comet relative to a star whose position is known.

Sector, zenith. An instrument for measuring the angular distance of a body from the *zenith*, generally when it is on the *meridian*.

Semi-diameter. With the Sun or Moon, it is generally a point on its *limb* that is actually observed, whereas what is needed is the measurement to its centre. Thus, the observed angle has to be corrected for semi-diameter, tabulated in the almanac.

Telescope, reflecting, familiarly a *reflector.* One which uses a figured mirror as its *objective*. The principal forms are *Cassegrain, Gregorian,* and *Newtonian.*

Telescope, refracting, familiarly a *refractor.* One which uses a lens as its *objective*. It may have either a simple or an *achromatic* objective.

Titius–Bode law. A numerical relationship between the mean distances of the planets from the Sun, first pointed out by J.B. Titius and popularized by J.E. Bode in 1772. When discovered, Uranus seemed to conform to the 'law', but Neptune and Pluto do not. It may be no more than a coincidence, but it is certainly very curious.

Transit in astronomy. 1. The passage of a celestial body (or a point like the *vernal equinox*) across the observer's *meridian*; 2. the passage of an apparently small body (like Mercury or Venus) across the disk of a larger one (like the Sun), the opposite to an *occultation*.

Transit clock. A clock in an observatory specially mounted for timing meridian *transits.*

Transit instrument. An instrument specially designed for observing the moment of *transit* of a body across the *meridian*.

Variation, magnetic. In nautical parlance, the amount by which magnetic north deviates from true north at any place, measured east and west of true north. To physicists, the same quantity is known as *magnetic deviation*.

Variation compass. A compass specially designed for measuring *magnetic variation*.

Zenith. The point directly overhead on the celestial sphere.

Zenith distance: the angular distance of a body from the *zenith*. Zenith distance is $90°$ minus *altitude*.

Notes

Abbreviations

Where only an author's name is cited, the full bibliographical reference will be found in the Bibliography. The following abbreviations are used in the references and in the Bibliography:

A/C	Maskelyne's account books; Arnold Forster MSS
AR	Astronomer Royal
BL	British Library, London
BM(NH)	British Museum (Natural History), London
DNB	*Dictionary of National Biography*
IOLR	India Office Library and Records, London
MB	Memorandum books of Nevil Maskelyne: WRO.1390/2/1–12
MS	Manuscript
nd	No date
NM	Nevil Maskelyne
NMM	National Maritime Museum, Greenwich
PRO	Public Record Office, London and Kew
RS	Royal Society, London
RGO	Royal Greenwich Observatory, Herstmonceux
RO	Royal Observatory, Greenwich
WRO	Wiltshire Record Office, Trowbridge

Preface

1. Flamsteed, J. (1675), *Historia Coelestis Britannica*, 3, London, 102. Translated by Mrs E.M. Barker.
2. Quill (1966).
3. Forbes (1975).

Chapter 1   The Maskelynes of Purton

1. NM autobiography, RGO MS.4/320:8, p.1. (in Appendix A below).
2. The Maskelyne family history here and in Appendix B has been obtained from three main sources: *Burke's Landed Gentry*, 15th ed. (1937), 1547–8; Mary Arnold-Forster's *Basset Down*; and the three Maskelyne volumes of G.F.M. Merriman's manuscript pedigrees (c.1908–10), BL MS Add.39690–2.

Notes to pp. 1–9

3. Margaret Lady Clive to Margaret Maskelyne, December 29 1802. Arnold-Forster Lady Clive MSS I.
4. Sainty, J.C.(comp.) (1973). *Office-holders in Modern Britain, ii – Officials of the Secretaries of State, 1660–1782*. London: HMSO, 88–9.
5. ibid.
6. 5 October is the date generally given for Nevil Maskelyne's birthday. However, in RGO.4/320/10, he himself clearly writes 6 October, whereas in other places, such as MB.2H/35^V he equally clearly writes 5 October. Britain did not adopt the Gregorian calendar (New Style) until September 1752.
7. Maskelyne, T.S. (1897), 127.
8. A few days before Edmund the father's death, Elizabeth Maskelyne asked the Duke of Newcastle for a clerkship in his office for Edmund, his second son (Elizabeth M to Newcastle, 7 March 1743/4; BL MS Add.32702, f.173). William Maskelyne's letter to Newcastle, of July 1753 (BL MS Add.32732, f.389), asking the latter to sponsor him as Professor of Hebrew, makes it clear that it was though Newcastle's influence that Edmund obtained the East India Company writership.
9. In fact, Flamsteed had used projected images of the sun at Greenwich from 1676. See the etching 'Domus Obscurata ad Maculas, Eclipsque Solares Excipiendas peropportuna', reproduced in Howse (1975b), plate Xa.
10. NM, op.cit. (ref.1).
11. *Phil.Mag.*, 42, 183 (July 1813), 3.
12. MB.2D/25^V.
13. Maskelyne, T.S. (1897), 132.
14. Lane Hall (1932), 14.
15. NM, op.cit. (ref.1), pp.1–2.
16. Bence-Jones (1974), 34. His very readable and well-documented *Clive of India* gives a great deal of information on the family life of the Clives and many of the references I cite were found originally because of that book.
17. Eliza Fowke to her aunts, 5 Feb. 1750/1. IOLR, Ormathwaite Collection, II.
18. Eliza Fowke to her aunt Fanny, 1 Aug. 1751. ibid.
19. NMM PST.76/234–5.
20. Margaret Maskelyne to Mrs. Casamajor, 7 December 1775. IOLR Fowke Collection, 21.
21. Bence-Jones (1974), 68–9.

Chapter 2 The longitude problem

1. There is a volume in the Wiltshire Record Office (WRO 1390/13) which contains a list of 'books bought or given me since my first coming to the University Novr.5.1749.' This continues to 1810 and appears complete, containing novels, Congreve's plays, etc. The cost of each book is noted.

2. RGO MS.4/207.
3. WRO.1390/16.
4. I am grateful to Mr Adam Perkins of the Royal Greenwich Observatory for drawing my attention to this gem.
5. RGO MS.4/340:8:10.
6. Private communication from the Archivist, Trinity College Library.
7. Bradley's Royal Warrant, RGO MS.4/320:3.
8. Act 12 Anne *c*.15, *An Act for Providing a Publick Reward for such Person or Persons as shall Discover the Longitude at Sea*.
9. Landes (1983), 112.
10. The full text of the Act is given in Quill (1966), 225–7.
11. For a fuller discussion on longitude-determination at sea, see Howse (1980), particularly 'Finding the Longitude', pp.192–8; and Forbes (1974)
12. Mayer (1770), cxvi.
13. Robertson, J. (1792), 308.
14. RGO MS.4/1.
15. *Gents. Mag.*, **28** (June 1758), 252–3.

Chapter 3 The transit of Venus, 1761

1. (a) Halley, E., De Visibili Conjunctione Inferiorum Planetam cum Sole, *Phil. Trans. R. Soc.*, **17** (1691), 511–22; &
   (b) Methodus singularis quâ Solis Parallaxis sive distantia à Terra, ope Veneris intra Solem conspiciendae, tuto determinari poterit, *Phil. Trans. R. Soc.*, **29** (1716), 454–64.
2. Quoted without source in Armitage, Angus (1966), *Edmond Halley*, London & Edinburgh, 104. For simple descriptions of this method of obtaining solar parallax, see Woolf (1959), Chapter I, and Woolley, Sir Richard (1969), Captain Cook and the Transit of Venus of 1769, *Notes and Records of the Royal Society*, **24** (1969), 19–32.
3. Woolf (1959) gives in his Chapter II a detailed account of French preparations and of Delisle's contribution.
4. *Gents.Mag.*, **28** (Aug.1758), 367–8 (letter from T. Fisher); **29** (Jan. 1759), 23–6 (summary of Le Gentil's papers from the *Mémoires* of the Académie des Sciences.); **30** (June 1760), 265–9 (a translation into English of Halley's 1716 Latin paper, see ref.1(b) above).
5. RS Council Minutes 3 July 1760 (all subsequent information on Council proceedings is from this source unless otherwise cited). RS Misc.MSS 10/106 reveals that a copy was sent to Madras in HMS *Chatham* well before the Royal Society first discussed the matter, while 10/150 gives Michell's name.
6. Woolf (1959), 211.
7. ibid., 59.
8. RS Journal Book, 19 June 1760.
9. NM (1761).
10. *Phil. Trans. R. Soc.*, **51** (1760), 865–88. Curiously, the date given in

Notes to pp. 24–29

the printed version for the reading of this paper is incorrectly given as 19 June. Furthermore, the minutes of the Ordinary Meeting of 26 June at which these two papers were read are omitted from the Society's Journal Book.
11. RS Council Minutes 3 July 1760 and Misc.MSS 10/100.
12. For the various Royal Society Clubs, see Allibone (1976). The dinner registers are preserved at the Royal Society.
13. Clevland to Birch, 30 July 1760. ibid.10/110.
14. Weld (1848), 14–15.
15. Smyth, W.H. (1860). *The Cycle of Celestial Objects continued at the Hartwell Observatory to 1859*, London. Four years later, Smyth's eldest son was to marry Maskelyne's grand-daughter, Antonia.
16. Woolf (1959), 86.
17. Monod-Cassidy (1980), 104. This agrees with the address given in Chapter 5, ref.2, both written later, in 1763.
18. In one of Dr Birch's small notebooks, there is an address written without explanation in Maskelyne's hand – 'Mr Waddington, at Matthew Pigott's Esqr., at Witton, near Twickenham'. This was presumably written in September 1760, giving Birch, one of the Society's Secretaries, Waddington's address, so that he could invite him to be an observer. BL MS Add.4471, 142.
19. Catalogue of instruments sent to Bencoolen. RS Misc.MSS 10/117a.
20. Mason to C. Morton, 8 December 1760; Dixon to Birch, 19 December; Mason to Macclesfield, 19 December. ibid.10/124; 10/125; 10/126.
21. NM to C. Morton, 27 December 1760. (The date of 27 Dec. is obviously a mistake and 27 Nov. is meant because this paper was presented to Council on Dec. 11) ibid.10/27. The instruments were described as follows: A 10-foot Astronomical Sector – 2 reflecting telescopes of two foot each (by Short) with a micrometer to one of them (by Dollond), a Helioscope, & an additional eyepiece with parallactic wires – A clock with a gridiron pendulum (by Shelton) – An equal altitude instrument – A stand for a Reflecting Telescope to observe near the Zenith – An alarum clock – A variation compass – A Dipping needle.

Chapter 4   Saint Helena, 1761
1. Mason to Morton, 12 Jan. 1761. RS Misc.MSS 10/128.
2. ibid.
3. Mason to Bradley, 25 Jan. 1761. ibid. 10/129.
4. Stephens to Birch, 31 Jan. 1761. ibid. 10/133.
5. RS Council Minutes 31 Jan. 1761.
6. Mason to C. Morton, May 6 1761. RS Misc.MSS 10/135.
7. Log of Lt. Wadham Osborne, HMS *Centaur*. NMM MS ADM/L/C.70.
8. RGO MS.4/150. At this date, when at sea (but not in harbour), ships' logs used the nautical day which *ended* at noon, so 4 p.m. 21st nautical date was the equivalent of 4 p.m. 20th civil date. To complicate matters further, nautical almanacs (and astronomers) used the astronomical

day which *started* at noon, and was thus 24 hours behind the nautical day. Happily, both the Royal Navy and the East India Company began to use the civil day in logs early in the nineteenth century. The astronomical day was not abandoned until 1925.
9. NM to Birch, 9 Sept. 1761. NM (1762e), 559.
10. NM to Birch, 13 May 1761. BL MS Add.4313, f.242.
11. Maskelyne (1763), 106.
12. NM op.cit. (ref.10), ff.240-1.
13. ibid. f.241, and RS Misc.MS 10/148 (reproduced in Appendix C).
14. ibid. f.240 and NM (1764c), 381.
15. Tatham, W.G. & Harwood, K.A. Astronomers and other scientists on St Helena, *Annals of Science*, **31**, 6, 503. A private communication from K.A. Harwood says that the figure 2013 in the above reference was published in 1904.
16. NM to Birch, nd [about 24 May 1761]. BL MS Add.4313, f.244.
17. NM to Lord Charles Cavendish, 30 July 1761. NM (1762d), 436.
18. Tatham & Harwood, op.cit.
19. C. Morton to Sullivan, Nov. 12 1761. RS Misc.MS 10/138.
20. NM op.cit. (ref.17).
21. Howse (1969), 282-4, 292-3.
22. NM to Macclesfield in NM (1762c), 196.
23. RGO MS.4/320:8, p.3 (in Appendix B below).
24. Woolf (1959) Chapter 4 and Woolley op.cit. (Chapter 3, ref.2) both give good summaries of the results of the transits of 1761 and 1768.
25. Morton op.cit. (ref.19).
26. Log of the *Oxford*. IOLR MS.
27. Morton op.cit. (ref.19).
28. NM (1764c). Much subsequent astronomical information is culled from the same source.
29. NM to Birch, 26 Jan. 1762 in NM (1762f). See also Cartwright, D.E. (1971-2), Some ocean tide measurements in the eighteenth century, and their relevance today, *Proc.R.S.E.(B)*, **72**, 32, 331-9.
30. NM, Observations with the zenith sector at St. Helena, 1761-2. RGO MS.4/2.
31. op.cit. (ref.23), p.5.
32. NM (1763), 113.
33. Log of the *Warwick*. IOLR MS.L/MAR/B/585H.

Chapter 5   The Barbados Trials, 1763-4

1. Most of the Harrison story told here and below has been culled from Quill (1966), which gives a well-researched, well-documented and most readable account. Even where other sources are cited, the clues as to where to look often came from Quill.
2. Mayer (1770), cxxvi.
3. Lalande (1759), *Connoissance des Temps pour l'Année 1761*, Paris, 174-93.

Notes to pp. 42–48

4. HEIC Court minutes 23 June 1762. IOL.B/78/70-1.
5. Waddington, Robert (1763). *A Practical Method for finding the Longitude & Latitude of a Ship at Sea by Observations of the Moon . . .*, London.
6. ibid. (1764). *A Supplement to the Treatise for finding the Longitude*, London.
7. John Robertson and Robert Waddington, respectively Mathematical Master and Usher at the Royal Naval Academy, Portsmouth, were dismissed in 1766, to be replaced by George Witchell and John Bradley (the astronomer's nephew), both of whom come into our story later.
8. RS Club dinner registers, vol.4.
9. Lady Clive to Edmund Maskelyne, 16 Dec. 1762. Arnold-Forster Lady Clive MSS.I.
10. RS Journal Book 9 June 1763. All subsequent quotations in this section come from a copy of this in RGO MS.6/21, 51-8.
11. NM to Birch, 9 Sept. 1761. NM (1762e), 559.
12. James Stewart Mackenzie MSS, kept with Bute papers, Mount Stewart Mackenzie MSS, kept with Bute papers, Mount Stewart, Rothesay. MS entitled 'Case of John Harrison, Inventor of the Instrument for finding Longitude at Sea', no author, no date, but February 1763. Contains statement, 'What seems to be the Wish of Every-Body, who knows any Thing of the Matter (Except Lord Macclesfield & Dr Bliss) is that the Principles . . . may be made known, by Parliament purchasing the same from Harrison . . .'. It seems likely that the writer was Taylor White.
13. Chapin, Seymour L. (1978), Lalande and the Longitude: a little known London Voyage of 1763, *Notes & Records of the R.Soc.*, **32** (1978), 171; and Monod-Cassidy (1980), 52.
14. *House of Commons Journal*, **29**, 515, 546-53.
15. Act 3 Geo. III *c*.14.
16. Rt.Hon. Lord Charles Cavendish, Rt.Hon. the Earl of Morton, Rt.Hon. Lord Willoughby of Parham, George Lewis Scott FRS, James Short FRS, Rev. John Mitchell (Michell), Woodwardian Professor of Geology at Cambridge, Alexander Cumming, Thomas Mudge, William Frodsham, Andrew Dickie, James Green.
17. Nivernois to Praslin, 21 March 1763; Praslin to Choiseul, 28 March 1763; Choiseul to Académie, 31 March 1763; Académie to Choiseul, 4 April 1763; extract from Académie register, 4 April 1763. Académie des Sciences, Paris.
18. Morton to Camus, 3 June 1763. J.S. Mackenzie papers, Rothesay.
19. Chapin op.cit. (ref.12), 61-2.
20. Monod-Cassidy (1980), 26.
21. RS Club Dinner Register 4.
22. Monod-Cassidy (1980), 26, 27, 29, 38, 48, 50, 78, 103.
23. These and subsequent proceedings of Board of Longitude meetings are taken from the appropriate volume of the Confirmed Minutes, RGO MS.14/5 (1737-79), 14/6 (1780-1801), 14/7 (1802-23).

24. NM to Edmund Maskelyne, 8 September 1763. NMM PST/76/ff.96-9.
25. Log of Lieutenant Patrick Fotheringham, H.M.S. *Princess Louisa*. NMM ADM/L/P.330.
26. Monod-Cassidy (1980), 32 & 106.
27. NM, 'Celestial Observations made on board his Majesty's Ship the Princess Louisa, Admiral Tyrrell, by Nevil Maskelyne. F.R.S.'. RGO MS.4/321:1.
28. NM to Edmund Maskelyne, 29 December 1763. NMM PST/76/f.103.
29. ibid. f.101.
30. NM to Bd. of Long., undated draft, probably 29 December 1763. RGO MS.4/320:7, p.2.
31. NM to Shepherd, 15 October 1764. NMM PST/76, f.131. See also Green's log in PRO ADM.51/4545/151.
32. NM, 'Astronomical observations made at the Island of Barbadoes.' RGO MS.4/323.
33. NM op.cit. (ref.28), ff.100-1.
34. NM op.cit. (ref.30), p.7.
35. Harrison MS Journal, 112-4. I am grateful to Mr. Andrew King for making a copy of this available to me.
36. Board minutes, 4 August 1763. RGO MS.14/5:27.
37. Quill (1966), 132.
38. NM, observations onboard the *Britannia*. RGO MS.4/1.
39. NM to Shepherd, 15 October 1764. NMM PST/76/130.

Chapter 6   Astronomer Royal, 1765

1. NM to Shepherd, 15 October 1764. NMM PST/76, ff.130-1.
2. Gould (1923), 63n.
3. NM to Shepherd, 20 October 1764. *The Observatory* 34, 441 (November 1911), 397. The three original letters transcribed are today missing from the RGO archives.
4. ibid. 396.
5. ibid. 396-7.
6. WRO 1390/3.
7. ibid.
8. RS Council Minutes 8 November 1764. Copy in RGO MS.6/21, 69-74.
9. *Gents. Mag.*, 34 (December 1764), 594-5.
10. Bevis had been elected a Member of Prussia's Royal Academy of Sciences in 1750. The 'Prussia hero' was Frederick the Great.
11. Bd. of Long. Confirmed Minutes 19 Jan. 1765. RGO MS.14/5:32.
12. Royal Warrant, 8 February 1765. WRO. 1390/3.
13. *Gents. Mag.*, 35 (February 1765), 95.
14. Royal Warrant, 22 February 1765. WRO. 1390/3; copy in RGO MS. 6/21, 81-4.
15. Sandwich to Granby, March 5 1765. RGO MS.6/21, 84-5.
16. Royal Warrant, 16 March 1765. RS MS.Gh.104 & WRO. 1390/3; copy in RGO MS.6/21, 75-7.

Notes to pp. 60–73   249

17. NM to Royal Society, 28 February 1765. RS MS.Gh.105.
18. ibid., 14 January 1766. RS MS.Gh.117.
19. RGO MS.3/14.
20. The Visitors comprised Lord Morton (PRS), James Burrow (VPRS), Ld. Charles Cavendish (VRPS), Capt. Campbell, Dr. Chandler, Dr. Heberden, Mr. Mauduit, Mr. West, Dr. Birch (Sec RS), Dr. Morton (Sec RS), attended by John Bird, instrument maker and John Shelton, clockmaker. RGO MS.6/21, 77–8.
21. RGO MS.6/21, 71.
22. Notes in Banks's hand, c1811. RS MS.Gh.130e^V. His Account Books show that the amount of fees and taxes on salary and pension did not change between 1773 and 1811.
23. Phelps (Sec. of State's Office) to E. of Morton, 6 February 1765, and NM to Jenkinson (Undersecretary of State), 14 February 1765. BL MS Add.38, 204, 64–5 & 74.

Chapter 7   The Royal Observatory and its instruments

1. Baily, F. (1835). *An Account of the Revd. John Flamsteed...* London, 37–8.
2. Flamsteed, J. (1725). *Historia Coelestis Britannica*, 3, 102, London. Translated here by Mrs E.M. Barker.
3. Baily op.cit, 112. Original in PRO SP.Dom.44, p.10.
4. PRO. WO.55/391/120.
5. Wren to Fell, 3 December 1681. *Wren Society*, 5, 21–2. Oxford: 1928.
6. For example, Forbes (1975) or Howse (1975b).
7. Flamsteed op.cit., 3 volumes.
8. Rigaud, S.P. (ed.) (1832). *Miscellaneous works and correspondence of James Bradley, D.D., Astronomer-Royal*. Oxford, lxxv.
9. Details of all these buildings and instruments are given in Howse (1975a).
10. NM to C. Morton, 28 March 1765. RGO MS.6/21, 86–9.
11. ibid., 13 May 1765. RS MS.Gh.111–2.
12. NM (1776), Transits, 26.
13. Bernoulli (1771), 87–8.
14. RGO MS.6/21, 88.
15. NM (1787b), Transits 277, Zenith distances 84.
16. NM (1799), Transits 383, Zenith distances 117.
17. ibid., Transits 397, Zenith distances 120.
18. NM (1776), Transits 221–47; NM (1811).
19. Evans, John (1810) (possibly Thomas's brother), *Juvenile Tourist*, London, 333–5.
20. MB.26/15^v & 2H/16^r.

Chapter 8 The Harrison affair, 1765-7

1. Guildhall MS.6026, quoted in Quill (1966), 136.
2. Board minutes, September 18 1764. RGO MS.14/5, 67. Subsequent references to the Board's proceedings in this chapter come from the same source and will not be cited except for quotations.
3. NM (1767a), p.(lvi), says that 9 emersions of the 1st satellite of Jupiter observed in Barbados, compared with 5 emersions observed by John Bradley at Portsmouth and 2 at Greenwich – with equal telescopes of two feet made by Mr. John Bird – gave a difference of meridians of 3h 54m 20s.
4. op.cit. (ref.2), 75.
5. ibid., 77-8.
6. ibid., 81.
7. Forbes (1971a), 6.
8. Harrison (1765).
9. Harrison MS Journal, 132.
10. NM to ?Shepherd, March 7 1765. *The Observatory*, **34**, 441 (November 1911), 393-6.
11. Harrison MS Journal, 164-6.
12. Quill (1966), 153.
13. Marguet (1931), 152.
14. Board minutes March 14 1767. RGO MS.14/5, p.146.
15. NM (1767a), 27. The window was bricked up when the Circle Room was built in 1809.
16. RGO MS.14/4/311. Quill, Fig.24, reproduces a page from this register.
17. NM (1767a), 5.
18. Harrison MS journal, 199-200.
19. ibid., 203-4.
20. Quill (1966), 165.
21. *House of Commons Journal*, **31**, 93.
22. NM (1767a), 24.
23. Harrison (1767a). This book is reprinted in facsimile in Betts (ed.) (1984).
24. op.cit. (ref.21), 270.
25. Gould (1923), 63n.
26. *Gentleman's Magazine*, **37** (Sept. 1767), 464-7, and Quill (1966), 173. Of the accusation that Maskelyne had a pecuniary interest in the rival method, *The European Magazine* for June 1805, p.408, says: 'This insinuation of interestedness in Dr. Maskelyne, all who knew him were sure was without foundation.'

Chapter 9 The Nautical Almanac

1. Board of Longitude minutes 28 May 1765. RGO MS.14/5, 88.
2. A/C 1, 5.
3. NM to Edmund Maskelyne, 15 May 1766. NMM PST/76, f.106.

Notes to pp. 86–107

4. 'Works Published by the Commiss: of Longitude', 1 January 1784. RS MS.MM.7.157. It is possible, though unlikely, that these figures do not include some copies supplied for official purposes.
5. Mayer (1770).
6. Cook in his journal, 23 August 1770. J.C. Beaglehole (ed.) (1968), *Voyage of the* Endeavour, *1768-1771*, Cambridge: Hakluyt Society, 392.
7. Sadler, D.H. (1976), Lunar distances and the Nautical Almanac, *Vistas in Astronomy*, **20** (1976), 113–121.
8. NM, in the Preface to every *Nautical Almanac* from 1767 to 1832. That for 1767 is reproduced in Sadler, 35–6.
9. Marguet (1931), 239–40 (in French).
10. RGO MS.4/231:1
11. R. Bishop (1773). *The East India Navigator's Daily Assistant*. London.
12. Op.cit. (ref.1), 12 November 1768, p.172.
13. Taylor (1956), 263.

Chapter 10 Early years at Greenwich, 1765-9

1. Royal warrant, 9 February 1765. WRO MS.1390/3.
2. NM to Edmund Maskelyne, 15 May 1766. NMM PST/76, pp.104–6.
3. Margaret Clive to Edmund Maskelyne, May 16 1766. Arnold-Forster MSS, Lady Clive letters, I.
4. NM, op.cit. (ref.2), p.106.
5. RS Club dinner register, vol. 5.
6. Private communication from Archivist, Trinity College Library, 29 October 1985.
7. RS Council minutes, 12 November 1767. RGO MS.6/21, 104.
8. RGO MS.4/3, p.121.
9. NM to Boscovich, 21 April 1767 & 12 December 1768. University of California, Berkeley, Bancroft Library MSS
10. Bernoulli (1771), 77–100.
11. Ewing to Franklin, 4 January 1769. Willcox, W.B. (ed.) (1973), *Franklin papers*, New Haven, **17**, 11.
12. NM (1769a), 355–65.
13. Woolf (1959), 182–7.
14. op.cit. (Chapter 9, ref. 4).
15. Bird (1767) & Bird (1768).
16. NM to Nourse, 14 April 1767. Hist. Soc. of Pennsylvania, Dreer collection 154.1, Astr. & Math. ii, 39–40.
17. Howse, Derek (ed.) (1986). The Greenwich List of Observatories, *J. Hist. astr*, vol. 17, part 4 (Nov. 1986).
18. Green to ?Dr. Morton, 11 September 1764. RS MS.Gh.101.
19. Dallaway to Dr. Morton, 22 October 1764. RS MS.Gh.102.
20. Board Minutes 12 December 1767. RGO MS.14/5, p. 166.
21. Dallaway to Dr. Morton, 4 February 1767. RS MS.Gh.119.
22. op.cit. (ref. 20) 4 March 1769. p.177.

23. ibid. 30 May 1765. p.93.
24. ibid. 14 December 1771. p.214.
25. NM to Howe, 16 February 1767. RGO MS.2/18, ff.25-6.
26. RS Council Minutes 24 October 1765.
27. Bill of lading, 6 December 1765. RS Gh.118. Royal Society Council minutes mixed up the name of the ship, the *Ellis*, with the name of her master, Egdon.
28. American Philosophical Society (1969). *The Journal of Charles Mason and Jeremiah Dixon*, Philadelphia, 211. Most of the other information in this section has been culled from this work.
29. Cook journal 29 January 1771. J.C. Beaglehole (ed.), (1955) *The Voyage of the Endeavour 1768-1771*, Cambridge, (new edition 1969), 448.

Chapter 11   The 1770s

1. Lady Clive, in a letter of 16 Sept. 1775 to NM, said of Miss Secker: 'I feel such a kindness for her (whether the child of her reputed father or of his first cousin JW) that I would allow you to make her from me a present of a hundred pounds.' Through Clive's influence, John Walsh was elected MP for Worcester in 1761, and would not have wanted it to be known that he had an illegitimate daughter.
2. Edmund Maskelyne to NM, 4 March 1773. NMM PST/76, 240-1.
3. ibid.
4. James Houblon Maskelyne to NM, 16 July 1774. NMM PST/76, 285-8. NM endorsed the letter: 'Recd. on Shehallion July 15th 1774'.
5. A/C I, 15, 16, 68.
6. Archivist, Trinity College, Cambridge op.cit. (Chapter 10, ref. 6).
7. A/C I, 49-50.
8. G.B. Airy (1855). *Address to the Visitors*, p.3.
9. The north quadrant and zenith sector were given achromatic object glasses in 1789 and 1801 respectively.
10. Royal Warrant 5 July 1771.
11. NM to Digby Marsh, 29 November 1790. *The Observatory*, 34, 441 (November 1911), 398.
12. NM to Ewing, 4 August 1775. American Philosophical Society Misc. MSS.
13. A/C I, 29-30.
14. The original manuscript is in Det kongelige Bibliotek in København, Ny kgl. saml. 377e (4to).
15. Shepherd (1772).
16. Mayer (1770).
17. Marguet (1931), 241.
18. Board minutes 12 November 1768, 17 November 1770, 2 March 1771. RGO MS.15/5, pp.174, 201, 203.
19. Cook & Green (1771). *Phil. Trans. R. Soc.*, 61, 397-421.
20. op.cit. (ref. 18), 28 November 1771, p.207.
21. ibid., p.209.

22. Solander to Banks, 5 September 1775. BM(NH) D.T.C. I, 98–99.
23. op.cit. (ref. 17), 28 November 1772, p.231.
24. Guildhall MS.6026 (no. 2), quoted in Quill (1966), 198–9.
25. *Parliamentary History of the House of Commons*, 17, (1813), cols. 812–3.
26. *House of Commons Journal*, 34, 244, 285–6, 298, 302, 367.
27. Forbes (1971a), 6.
28. NM autobiographical notes. RGO MS.4/320:10, f.2.
29. Cook to Stephens, 22 March 1775. J.C. Beaglehole (ed.) (1959). *The voyage of the Resolution and Adventure 1772–1775*, Cambridge: Hakluyt Society, 692.
30. op.cit. (ref.18), 27 May 1775, p.276.,

Chapter 12   Weighing the world – Schiehallion, 1774

1. NM (1768b), 273.
2. NM (1768d), 328.
3. Lord Brougham, quoted in *DNB*, p.1261.
4. NM (1775a).
5. ibid., 496.
6. NM to James Lind, August 23 1773. RS MS.244/12.
7. NM (1775b), 502–3.
8. ibid., 505.
9. ibid., 500–42. Leadstone (1974) and Davies (1985) give fuller non-technical accounts.
10. Most of the information for this anecdote came from papers kept by Mr Duncan Robertson of Rannoch (a descendant of the Duncan in the anecdote), the only reference I have found in print, very brief, being in Alexander Stewart (1928), *A Highland Parish or the History of Fortingall*, Glasgow, p.67. The Gaelic song is published in James Robertson's Memories of Rannoch, *Transactions of the Gaelic Society of Inverness*, 51 (1978–80), 317–9. The translation here is by Alison Philips of Killichonan. For the subsequent history of the fiddle, I am indebted to Professor J.F. Allen of St. Andrews.
11. NM to Rev. Digby Marsh, November 29 1790. Op.cit. (Chapter 11, ref. 10), 397–8. Maskelyne was paid £212.0.11 expenses: the whole experiment cost the society (out of the King's grant) £597.16.0, including £150 to Hutton for his analysis.
12. NM (1775b), 532–4.
13. Pringle, Sir John (1775). *A Discourse on the Attraction of Mountains, delivered at the Anniversary Meeting of the Royal Society November 30 1775 by Sir John Pringle, Baronet, President.*, London, 31.
14. Burrow's journal no. 1, 10 September 1775. RAS MS.
15. de Morgan, Augustus (1864). *Notes and Queries*, 3rd series, 5, 361.
16. Burrow to Royal Society, 26 January 1775. RS MS.M.M.4:81.
17. *St. James's Chronicle or British Evening-Post*, no. 2327, 11–13 January 1776, p.2.

254   Notes to pp. 141–155

18. Hutton (1778), 766. These methods are presumably those described in [Cavendish, Henry], 'Mr. Cavendish's rules for computing the attraction of Mountains on Plumblines': two similar papers, RGO MS. 4/115 & 116.
19. ibid., 782.
20. ibid., 783.

Chapter 13   The 1780s – and a new planet

1. A/C I/86.
2. *Gentleman's Magazine*, **82**, ii (Sept. 1812), 220–1.
3. Probate of will of Mrs Martha Henn Rose dated January 23 1784. Northampton Record Office C(S)327/10.
4. MB.2D/19.
5. A/C II/2. Other parts of the story canbe pieced together from account book entries.
6. For many of my facts in this section, I have relied upon Simon Schaffer's excellent paper 'Uranus and the establishment of Herschel's astronomy', *J.Hist.astron.* (1981), **12**, 11–26.
7. NM to Watson Junior, 4 April 1781. RAS Herschel MSS. W1/13.M.14.
8. NM to Banks, 16 April 1781. NMM MS.PST/76, 146–8.
9. Herschel to Lalande, September 1782. Op.cit. (ref. 6), 15.
10. RS Journal Book, **30**, 408–12.
11. William to Caroline Herschel, 3 June 1782. Mrs John Herschel (1876), *Memoir and Correspondence of Caroline Herschel*, London, 47.
12. NM to Herschel, 8 August 1782. RAS Herschel MSS.W1/13.M.20.
13. Bugge to NM, 2 July 1784. København, det Kongelige Bibliotek, letter book pp.36–8.
14. NM to N. Pigott, 4 September 1783. RAS Pigott MS 50.
15. British Museum, Dept. of Prints and Drawings 1872-7-13-481.
16. Blagden to Banks, 21 October 1783. BM(NH).DTC.3, 139–42 (Dawson, 60).
17. Blagden to Banks, 1783. Quoted in Alibone (1976), p.120.
18. *Phil. Trans. R. Soc.* (1784), **74**, 201ff.
19. ibid. (1790), **80**, 134ff.
20. This whole episode is told in detail by Forbes (1975), pp.147–50, and in more detail still in his The Geodetic Link between the Greenwich and Paris Observatories in 1787, *Vistas in Astronomy* (1985), **28**, 173–81.
21. Memo Book no. 5. RGO MS.4/327.
22. NM (1787a), 186.
23. MB.2D/32^V.
24. Roy, Major-general William (1790), An Account of the Trigonometrical Operation, whereby the Distance between the Meridians of the Royal Observatories of Greenwich and Paris has been determined. *Phil. Trans. R. Soc.*, **80**, 111ff.
25. MB.2D/30^V.
26. MB.2D/31.

27. MB.2F/16, 6/18.
28. NM to William Herschel, October 25 1786. RAS Herschel MS.13.M.30.
29. MB.2E.
30. MB.2D/34^V.
31. In the Preface, Maskelyne claims that stocks of the first edition were exhausted. He was mistaken because 6000 unsold copies were sent for pulping in 1800 (see p.198).
32. NM (1786).
33. Hunter, J. (1793). *An Historical Journal . . . from the first sailing of of the Sirius in 1787, to the return . . . in 1792.* London, 32.
34. Dawes to NM, various letters. RGO MS.14/36, 239–308.
35. *Historical Records of Australia*, series 1, vol.1, 676 & 701. I am grateful to Dr Alan Day of Sydney for drawing my attention to this.
36. Anon. (1784). *An Authentic Narrative of the Dissensions & Debates in the Royal Society, containing speeches at large of Dr. Horsley, Dr. Maskelyne, Mr. Maseres, Mr. Poore, Mr. Glenie, Mr. Watson and Mr. Maty.*, London.
37. Cameron (1952), 130.
38. Blagden to Banks, December 27 1783. BM(NH) DTC.3.180-1.
39. Cameron (1952), 131.
40. MB.4/13.
41. ibid.
42. Several pamphlets published in 1784 concerning the dissensions are bound in the Royal Society's volume Tracts 3: most are against Banks (Horsley, Maty and others), one is for him (Andrew Kippis). For modern accounts, see Cameron (1952), 128–45; and Miller, David Philip (1983), Between Hostile Camps, *Brit. J. Hist. Science*, 16, 1–47: both are well documented.
43. Anon. [Olinthus Gregory] (1820). A review of some leading Points in the Official Character of the late President of the Royal Society, *Phil. Mag.* (Sept. 1820), 161–74
44. See Cameron (1952), 252.
45. Blagden to Banks, October 24 1784. BM(NH) DTC.4, 85–6.
46. NM to Herschel, December 19 1791. RAS W1/13.M.51.

Chapter 14   The 1790s

1. MB.2G/12.
2. MB.2H/22^V.
3. This line of descent is shown on a sheet of notes made by Thereza Story-Maskelyne, a copy of which is in the possession of Mrs Vanda Morton. See the family tree in Appendix B.
4. MB.2H/39.
5. Ryan, W.F. (1966), John Russell, R.A., and early lunar mapping, *The Smithsonian Journal of History*, 1, 27–48. As well as those mentioned by Ryan, one of Russell's moon pastels and a Selenographia are preserved at the National Maritime Museum, Greenwich. On Russell, see

also Williamson, G.C. (1894), *John Russell, R.A.* London.
6. Rees, *Cyclopaedia*, **22**, article 'Maskelyne'.
7. RS Council minutes 8 August 1792.
8. *DNB*, 16, 709.
9. NM (1799), 339.
10. Brooks, G.C., & Brooks, R.C. (1979). The improbable progenitor, *J.Roy. astron.Soc.Can.*, **73**, 1, 9–23.
11. MB.2H/6.
12. Delambre (1813), 14.
13. Hoefer (1854). *Nouvelle Biographie Universelle*, **9**, col. 526.
14. Board minutes, 14 August 1784. RGO MS.14/6, p.71.
15. For Zach's career and particularly his and Brühl's part in the finding and attempted publication of the scientific papers of Thomas Harriot, the Elizabethan navigator and astronomer, see Shirley, John W. (1983), *Thomas Harriot: a biography*, Oxford, 14–26. Shirley's main source is a rare pamphlet by Rigaud, S.P. (1833), *Supplement to Dr. Bradley's Miscellaneous Works: with an account of Harriot's astronomical papers*, Oxford, which explains 'why Oxford refused to publish Zach's work on the subject'. See also Hutton, Charles (1796), 584–6.
16. op.cit. (ref. 14), 160.
17. Gould (1923) in his Appendix I (pp.253–66) discusses, lucidly and in detail, the various methods of chronometer rating. He concludes that Maskelyne's method can be criticised in that, if a chronometer's performance improves during the course of the trial, it will appear to be getting worse, and the longer the trial the worse it will appear. Nevertheless, the other methods suggested at the time were even less satisfactory.
18. *House of Commons Journal*, **47**, 521.
19. Banks to Windham, 12 March 1793. RS MM.7.117.
20. Shuckburgh to Banks, 24 April 1792. RS MM.7.89.
21. Banks to Shuckburgh, 25 April 1792. RS MM.7.88.
22. Gould (1923) notes (p.79) that two of Maskelyne's sentences contain respectively 150 and 302 words.
23. NM (1792a), 126.
24. *Report from the Select Committee of the House of Commons to whom it was referred to Consider of the Report which was made from the Committee to whom the Petition of Thomas Mudge watch-maker was referred*, (1793), 82.
25. For another account of the history of this complicated affair, see Cameron (1952), 235–8; for the technical aspects, see Gould (1923), 76–82.
26. Board minutes 7 December 1793. RGO MS.14/6, 207–8.
27. The tables used for calculating the ephemeris for that particular year were stated in the Preface to each almanac.
28. *Lois, Décrets, Ordonnances et Décisions concernant le Bureau des Longitudes*, Paris, 1909, 1–15.
29. NM to Banks, 8 July 1779. RS MM.7.11.

Notes to pp. 178–192

30. Rees (1812), op.cit.
31. MB.2G/22^V.
31. A/C 2/49, 3/71.
33. Gooch to his parents, 21 April 1791. Cambridge University Library MM.6.48, f.26.
34. NM to William Gooch, senior, 26 November 1793. ibid., f.108.
35. ibid., 14 January 1794. ibid. f.114.
36. Maskelyne Point, on Maskelyne Island, at the entrance to Portland Inlet, in 54[39′N], 130[27′W].
37. For example, Banks to Hornsby, 19 November 1784 (Kew: B.C.1.180), and Banks to Price, 16 November 1785 (Bodley MS Add.A.64.20–21).
38. The source of most of the information in this section is contained in the 22-page pamphlet, *Proceedings of the Board of Longitude For the Recovery of the late Dr. Bradley's Observations, with some other papers relative thereto. 6th June 1795.*, a copy of which is in RGO MS.14/4, ff.260–70. See also Forbes (1965).
39. Portland to Spencer, 6 June 1795. RGO MS.14/4, f.270.
40. op.cit. (ref. 35).
41. NM to Banks, 28 March 1798. NMM PST/76, 182–3.
42. MB.2H/12.

Chapter 15   The final years, 1800–11

1. *Reports from Committees of the House of Commons*, 14 (1803), 606–8. See also Skempton, A.W. (1980). Telford and the Design for a new London Bridge, *Thomas Telford, Engineer. Proceedings of a seminar... April 1979*, London: Thomas Telford Ltd. There are several Maskelyne letters in the Institute of Civil Engineers MSS. T/LO.
2. RGO MS.4/129, iv.
3. NM, no address, no date but summer 1801. NMM PST/76, 117–8.
4. Most of the information in this section is culled from Forbes (1971b) which gives a detailed account of these episodes. See also Forbes, E.G. (1971c), Gauss and the discovery of Ceres, *J.Hist.astron.*, 2, 195–9.
5. NM to ?Banks, 14 May 1802. NMM PST/76/165–6.
6. Zach to NM, 4 May 1802. RGO MS.4/119/x.
7. NM to ?Banks, 23 January 1801. NMM PST/76/173–4.
8. Inman to NM, 5 August 1803. RGO MS.14/54, 209ff.
9. MB.2F/15^V.
10. Except where otherwise indicated, all the information in this section is culled either from the Board of Longitude minutes (RGO MS.14/6 or 14/7) or from Earnshaw (1808). For a technical discussion see Gould (1923), 116–28, and Betts (1984).
11. Maskelyne (1804), 1.
12. ibid., 13–14.
13. Earnshaw (1808), 189.
14. Banks to NM, 24 February 1806. BM(NH). DTC.16, 236–7.

15. NM to Banks, 24 February 1806. ibid., 235.
16. Dampier to Banks, 3 March 1806. ibid., 243–8.
17. Dalrymple (1806). For relations between Dalrymple and Arnold, see Cook (1985), 189–95.
18. Earnshaw (1808).
19. NM to Banks, 26 May 1809. RS Misc. MS.8.67.
20. *The Times*, 18 November 1801, 2.
21. NM to Margaret Maskelyne, 20 November 1801. NMM PST/76, 109–110.
22. A/C 3/5.
23. MB.2J/25^V.
24. *The Times*, 4 June 1805, 2.
25. Memorial in NM's hand. RS MS.Gh.75–6, no date but presented at Visitation of 11 July 1806.
26. Pond, J. (1806), On the Declinations of some of the principal fixed stars, *Phil.Trans. R. Soc.*, **96**, 420–54.
27. MB.2K/4.
28. RS Council Minutes 28 May 1807 & 24 March 1808.
29. Delambre to NM, 20 February 1806. NMM PST/76, 75–7.
30. Hardy to ?Margaret Maskelyne, 7 June 1820. NMM PST/76, 125.
31. Board minutes 7 December 1809. RGO MS.14/7, 134.
32. Draft in NM's hand, c,1809. RS MS.Gh.158.
33. Banks to NM, 8 January 1810. RS MS.Gh.137.
34. RS Council Minutes 29 March 1810.
35. Royal Warrant 1 May 1810. WRO.1390/3.
36. Banks to the Treasury, 1814. RS MS.Gh.12.
37. NM's will, 25 October 1810. WRO 1390/94; transcription in WRO 1390/123.
38. NM to Andrews, 5 January 1811. RGO MS.4/149, 50.
39. NM to Vince, 5 January 1811. NMM PST/76, 121–4.
40. Note by Margaret Maskelyne. ibid, 119.
41. Margaret Maskelyne to Andrews, 7 February 1811. RGO MS.4/149, 51.
42. Margaret Maskelyne to Lady Booth, 11 February 1811. NMM PST/76, 281–4.
43. NM op.cit. (ref. 37).
44. A.M. Clerke, *DNB*, article 'John Pond'.
45. Pond, receipt for books & MSS, & certificate that instruments appear to be in good order, 13 April 1811. NMM PST/76, 1–10.
46. For gross sum, see annotated sale catalogue, WRO 1390/12; for net sum, A/C 3/118.
47. Anon. [Gregory] op.cit. (Chapter 13 above, ref. 43), 247.

Chapter 16   Summing up

1. Rees, *Cyclopaedia*, **22**, article 'Maskelyne'.
2. ibid.
3. Delambre (1813), translation in *Phil.Mag.* **42**, 1 July 1813), 141.

Appendix B Nevil Maskelyne's autobiographical notes

1. See Chapter 1, note 9.
2. According to university records, Maskelyne was Seventh Wrangler, whereas he implies here that he was Third. The cause of this discrepancy is not known.
3. The Governor in Halley's time was Gregory Field. He was dismissed in 1678 after numerous complaints.
4. Lord Morton died in 1768.
5. Thomas Hornsby, D.D. (1733–1810).
6. John Smith (d.1797). His successor in the Geometry chair, Abram Robertson, completed the edition of Bradley's observations, the second volume being published in 1805.
7. The text as eventually edited in another hand (probably Margaret Maskelyne's; her words in square brackets) reads:

   However it was not till [the year 1798] that the first Volume of it in Folio was published, [not until] *The learned world [have since been gratified by] the publication of the [whole under the superintendence of Dr. Robertson the Savilian Professor of Geometry at Oxford &c &c] We flatter ourselves that our readers will not be displeased with our giving the continuance of the history of the observations, so interesting to Astronomy, the first steps for the recovery & publication of which were made by [Dr. Maskelyne].

8. The last six words were heavily scored out by the editor and replaced by, 'The merits of Dr. Maskelyne alone'.

Bibliography

NM = Nevil Maskelyne. For other abbreviations, see p.242.

Principal manuscript sources

Nigel Arnold-Forster Esq.,
Basset Down, nr. Swindon, Wilts SN4 9QP
96 letters to and from NM, cited as NMM MS PST/76.
  Photostats at NMM and RGO.
Three account books, cited as A/C I, II, III, Microfilms at NMM.
199 letters from Margaret Lady Clive.
  Photocopies at IOLR, no.Photo Eur.287.
A few miscellaneous papers.

British Library, London (BL)
MS Add.39, 690-2 – Merriman Pedigrees, 7 – Some Account of the Family of Maskelyne of Purton in the County of Wilts. . . Microfilm at NMM.

National Maritime Museum, Greenwich (NMM)
MS PST/76 – photostats of Arnold-Forster letter collection.
MS MRF/152 – microfilms of Arnold-Forster Account Books.
MS MRS/153 – microfilm of WRO Memorandum Books
MS MRS/ – microfilm of BL Merriman Pedigrees, 7
MS XBLN/1-56 – photocopies of RGO MS.14, Board of Longitude papers.
MS PST/F/2-3 – photocopies of RGO MS 6/21 and 6/22, RS Council minutes concerning RO.
MS PST/ – photocopies of Rs MS.class Gh – papers about RO.

Royal Greenwich Observatory, Herstmonceux (RGO)
RGO MS 4 – Papers of Nevil Maskelyne
RGO MS 6/21, 6/22 – Transcription of RS Council Minutes concerning the Royal Observatory
RGO MS 14 – Papers of the Board of Longitude. These contain many letters to and from N.M. His accounts with the Board are in MS 14/18, ff.16-130.
RGO MS 35:1-96 – Photostats of Arnold-Forster letter collection.

Royal Society, London (RS)
RS MS vols. 371, 372 (class Gh) – Papers concerning the Royal Observatory.

Wiltshire Record Office, Trowbridge
WRO 1390 – Arnold-Forster papers, particularly /2, /3, /10, /12, /13, /25, /49, /94, /96, /98, /107, /116, /123.
WRO 1390/2 – Twelve pocket memorandum books, cited as MB.2A, 2B, etc. Microfilm at NMM.

Nevil Maskelyne's published works

Prices of Board of Longitude items as at 1814.

NM (1761). A Proposal for discovering the annual Parallax of Sirius. *Phil.Trans.R.Soc.* 51, 889–95.

– (1762a). A Theorem on the Aberration of the Rays of Light refracted through a Lens on account of the Imperfection of the Spherical Figure. *Phil.Trans.R.Soc.* 52, 17–21.

– (1762b). Observations to be made at St. Helena, to settle differences of Longitude, 7c. [from Lacaille, with reply by NM]. *ibid.*, 21–7.

– (1762c). Account of the Observations made on the Transit of Venus, June 6, 1761, in the Island of St Helena. *ibid.*, 196–201.

– (1762d). Observations on a Clock of Mr. John Skelton (*sic*), made at St Helena. *ibid.*, 434–43.

– (1762e). Results of Observations of the Distance of the Moon from the Sun and fixed Stars, made in a Voyage from England to St Helena, in order to determine the Longitude of the Ship from time to time, together with the whole Process of Computation used on this Occasion. *ibid.*, 558–77.

– (1762f). Observations of the Tides in the Island of St Helena. *ibid.*, 586–606.

– (1763). *The British Mariner's Guide containing Complete and Easy Instructions for the Discovery of the Longitude at Sea and Land, within a Degree, by Observations of the Distance of the Moon from the Sun and Stars, taken with Hadley's Quadrant.* London: for the author.

 – (1764a). Concise rules for computing the effects of Refraction and Parallax in varying the apparent distance of the Moon from the Sun or a Star, also an easy Rule of Approximation for computing the Distance of the Moon from a Star, the Longitudes and latitudes of both being given, with Demonstration of the same. *Phil.Trans.R.Soc.* 54, 263–76.

 – (1764b). Some remarks upon the Equation of Time and the true Manner of computing it. *ibid.*, 336–47. Latin text in *Acta Eruditorum* (Leipzig) (1764).

 – (1764c). Astronomical Observations made at the Island of St Helena. *Phil.Trans.R.Soc.* 54, 348–86.

 – (1764d). Astronomical Observations made at the Island of Barbadoes; at Willoughby Fort; and at the Observatory on Constitution Hill, both adjoining to Bridge Town. *ibid.*, 389–92.

 – (ed.) (1766a). *The Nautical Almanac and Astronomical Ephemeris for the year 1767.*, London: Commissioners of Longitude. 3s.6d.

- (ed.) (1766b). *Tables Requisite to be used with the Astronomical and Nautical ephemeris.* London: Commissioners of Longitude. 2s.6d.
- (ed.) (1767–1810). *The Nautical Almanac and Astronomical Ephemeris.* . . NM edited 49 almanacs, for the years 1767 to 1815. London: Commissioners of Longitude. In sheets, 1767–1801 3s.6d., 1801–1814 5s.
- (1767a). *An account of the Going of Mr. John Harrison's Watch, at the Royal Observatory, from May 6th, 1766 to March 4th, 1767.* . . London: Commissioners of Longitude. 2s.6d.
- (1767b), *see* Harrison, John (1767a).
- (1768a). Introduction to two papers of Mr. Smeaton [on menstrual parallax and observations out of the meridian]. *Phil.Trans.R.Soc.* **58**, 154–5.
- (1768b). Introduction to the following Observations, made by Messieurs Charles Mason and Jeremiah Dixon, for determining the Length of a Degree of Latitude, in the Provinces of Maryland and Pennsylvania in North America. *ibid.*, 270–3.
- (1768c). The length of a Degree of Latitude in the Provinces of Maryland and Pennsylvania deduced from the foregoing operations [by Mason and Dixon]. *ibid.*, 323.
- (1768d). Postscript by the Astronomer Royal [observations on proportion of English and French measures]. *ibid.*, 325–8.
- (1768e). Instructions relative to the observation of the Ensuing Transit of the Planet Venus over the Sun's Disk on the 3rd of June 1769. Appendix to *The Nautical Almanac . . . for the year 1769.* London: Commissoners of Longitude.
- (1769a). Remarks on Wright's observations of the transit of Venus at Isle Coudre. *Phil.Trans.R.Soc.*, **59**, 279.
- (1769b). Observations of the transit of Venus over the Sun, and the Eclipse of the Sun, on June 3, 1769; made at the Royal Observatory. *ibid.*, 355–65. *Also published in Amer.Phil.Soc.Trans.*, **1** (1771), 1–4.
- (1769c). Eclipses of Jupiter's First Satellite, the eclipse of the Moon, and Occultations of Fixed Stars observed at the Royal Observatory at Greenwich, in the year 1769. *Phil.Trans.R.Soc.* **59**, 399–401.
- (1769d). Remarks on Biddle and Bayley's observations of the transit of Venus. *ibid.*, 420.
- (1770a). A Correct & Easy Method of Clearing the Apparent Distance from the Moon from a Star or the Sun of the effects of Refraction & Parallax by the help of three tables. Appendix to *The Nautical Almanac . . . for the year 1772.* London: Commissioners of Longitude.
- (ed.) (1770b), *see* Mayer, Tobias (1770).
- (1771a). A Catalogue of the Places of 387 Fixed Stars . . . calculated from the late Dr. Bradley's Observations. Appendix to *The Nautical Almanac . . . for the year 1773.* London: Commissioners of Longitude.
- (1771b). Description of a Method of measuring Differences of Right Ascension and Declination, with Dollond's Micrometer, together with other new Applications of the same. *Phil.Trans.R.Soc.* **61**, 536–46.
- (1772a). Directions for using a common micrometer, taken from a paper in the late Dr. Bradley's handwriting. *Phil.Trans.R.Soc.* **62**, 46.
- (1772b). Remarks on the Hadley's Quadrant, tending principally to remove the Difficulties which have hitherto attended the Use of the Back-observation, and to

obviate the Errors that might arise from a Want of Parallelism in the two Surfaces of the Index-glass. *ibid.*, 99–122. Also printed as an appendix to *The Nautical Almanac . . . for the year 1774*. London: Commissioners of Longitude.

- (1774a). M. de Luc's Rule for measuring heights by Barometer, reduced to the English measure of length and adapted to Fahrenheit's Thermometer, and other scales of heat, and reduced to a more convenient expression. *Phil.Trans.R.Soc.* 64, 158–70.
- (1774b). Observations of the Eclipses of Jupiter's first Satellite made at the Royal Observatory at Greenwich, compared with observations of the same, made by Samuel Holland Esquire, Surveyor General of Lands for the Northern District of America, and others of his Party in several parts of North America and the longitudes of the places thence deduced. *ibid.*, 184–93.
- (1774c). *Tables for computing the apparent Places of the Fixt Stars, and reducing Observations of the Planets*. London: Royal Society at public expense.
- (1775a). A Proposal for measuring the Attraction of some Hill in this Kingdom by astronomical Observations. *Phil.Trans.R.Soc.* 65, 495–99. (Read at RS 1772).
- (1775b). An Account of Observations made on the Mountain Schehallien, for finding its Attraction. *ibid.*, 500–42.
- (1776). *Astronomical Observations made at the Royal Observatory at Greenwich from the year MDCCLXV to the year MDCCLXXIV. Volume I*. London: President and Council of the Royal Society at Public Expense in obedience to his Majesty's Command.
- (1777). Account of a new Instrument for measuring small angles, called the Prismatic Micrometer. *Phil.Trans.R.Soc.* 67, 799–815.
- (1779). On the Longitude of Cork. *Phil.Trans.R.Soc.* 69, 179–81.
- (ed.) (1781). *Tables Requisiite to be used with the Nautical Ephemeris for finding latitude and Longitude. Second edition corrected and improved.* London: Commissioners of Longitude. 5s.
- (1783a). Remarks on the Tremors peculiar to Reflecting Telescopes. Appendix to *The Nautical Almanac . . . for the year 1787*. London: Commissioners of Longitude.
- (1783b). *A Plan for observing Meteors called Fire-balls*. Greenwich.
- (1786). Advertisement of the expected return of the Comet of 1532 and 1661, int he year 1788. *Phil.Trans R.Soc.*, 76, 426–31.
- (1787a). Concerning the Latitude and Longitude of the Royal Observatory at Greenwich; with remarks on a Memorial of the late M. Cassini de Thury. *Phil.Trans.R.Soc.* 77, 151–87.
- (1787b). *Astronomical Observations . . . from the year MDCCLXXV to the year MDCCLXXXVI. Volume II.* London: Royal Society at public expense.
- (1789). An attempt to explain a Difficulty in the Theory of Vision, depending on the different Refrangibility of Light. *Phil.Trans.R.Soc.* 79, 256–64.
- (1792a). *An Answer to a Pamphlet entitled 'A Narrative of the Facts' lately published by Mr Thomas Mudge, Junior, relating to some Time-keepers constructed by his father Mr Thomas Mudge; wherein is given An Account of the Trial of his first Time-keeper and of the three Trials of his other Time-keepers, between the years of 1774 and 1790, by Order of the Board of Longitude at the Royal Observatory. And also the Conduct of the Astronomer Royal, and the Resolutions of the Board of Longitude, respecting them, are vindicated from Mr Mudge's misrepresentations.* London: Commissioners of Longitude.
- (1792b). Preface and Precepts for the Explanation and Use of Taylor's Table of Logarithms, *see* Taylor, Michael (1792).

- (1793). Observations on the Comet of 1793. *Phil.Trans.R.Soc.* **83**, 55.
- (1794). An Account of an Appearance of Light, like a Star, seen lately in the dark part of the Moon, by Thomas Stretton, in St. John's Square, Clerkenwell, London; with Remarks upon this Observation, and Mr. Wilkins's. *Phil.Trans.R.Soc.* **84**, 435.
- (1799). *Astronomical Observations . . . from the year MDCCLXXXVII to the year MDCCXCVIII. Volume III.* London: Royal Society at public expense.
- (ed.) (1802). *Tables Requisite . . . 3rd edition.* London: Commissioners of Longitude. 5s.
- (1804). Arguments for giving a Reward to Mr. Earnshaw for *his improvements on Time-keepers.* Greenwich.
- (ed.) (1806). *Explanations of Time-keepers constructed by Mr. Thomas Earnshaw and the late Mr. John Arnold.* London: Commissioners of Longitude. Reprinted in Betts, Jonathan (1984).
- (1807). On the property of the tangent of three arches trisecting the circumference of a circle. *Phil.Trans.R.Soc.* 98, 122–3. Also published in W. Nicholson's *Jnl.Nat.Phil.Chem. & Arts*, **98** (1808), 340.
- (1811). *Astronomical Observations . . . from the year MDCCXCIX to the year MDCCCXI. Volume IV.* London: Royal Society at Public Expense.

Other primary published works

Arnold, John
- (1782). *An Answer from John Arnold to an Anonymous Letter.* London.
- (1791). *Certificates and Circumstances relating to the going of Mr. Arnold's Chronometers.* London: J. Ramshaw.

Arnold, John Roger (1805). *Explanation of Time-keepers. . .*, see NM (ed.) (1806).

Banks, Joseph (1804). *Protest against a Vote of the Board of Longitude, granting to Mr. Earnshaw a Reward for the Merit of his Time-keepers.*

Bernoulli, Jean (1771). *Lettres Astronomiques*, Berlin, pp.77–100.

Bird, John
- (1767). *The Method of Dividing Astronomical Instruments.* London: Commissioners of Longitude.
- (1768). *The Method of Construction of mural quadrants, exemplified by a description of the brass mural quadrant at the Royal Observatory at Greenwich.* London: Commissioners of Longitude.

Dalrymple, Alexander (1806). *Longitude. A full answer to the advertisement concerning Mr. Earnshaw's Timekeeper in the* Morning Chronicle, *4th Feb. and* Times *13th Feb., 1806.* London: for the author.

Delambre, J.B. (1813). *Notice sur la Vie et les Travaux de M. Maskelyne, lue à la Séance publique de l'Institut National de France, du 4 janvier, 1813, par M. le Chevalier Delambre, Sécretaire Perpétuel,* London: Hansard Jr. An English translation was published in *The Philosophical Magazine*, 42, 183 (July 1813).

Earnshaw, Thomas
- (1804). *An Explanation of the Escapement of Mr. Earnshaw's Time-keeper – from a model presented to a meeting of the Board of Longitude, 7th of June 1804.*
- (1805). *Explanation of Time-keepers. . .* see NM (ed.) (1806).
- (1808). *Longitude. An Appeal to the Public: stating Mr. Earnshaw's claim to the original*

invention of the improvements in his timekeepers, their superior going in numerous voyages, and also as tried by the Astronomer Royal by orders of the Commissioners of Longitude, and his consequent right to National Reward. London: for the author.

Harrison, John
- (1763). *An account of the Proceedings...*, see Royal Society, a member of.
- (1765). *A Narrative of the Proceedings relative to the Discovery of the Longitude at Sea.* London.
- (1767a). *The principles of Mr. Harrison's Time-keeper, with plates of the same,* London: Commissioners of Longitude. 5s. French edition, E. Pézénas (tr. & ed.) Paris and Avignon, 1767. Reprinted by British Horological Institute, 1984, with Introduction and Technical Appraisal by Betts, J.
- (1767b). *Remarks on a Pamphlet lately published by the Rev. Mr. Maskelyne, under the Authority of the Board of Longitude.* London: for the author.

Hutton, Charles
- (1778). An Account of the Calculations made from the Survey and measures taken at Schehallien, in order to ascertain the Mean Density of the Earth, *Phil.Trans.R.Soc.*, **68**, 689–788.
- (1796). *A Mathematical and Philosophical Dictionary...*, 2 vols., London.
- (1815). *A Philosophical and Mathematical Dictionary...*, New edition, 2 vols. London, 2, 22–3, article 'Maskelyne'.

Lalande, J.-J. Lefrançais
- (1763), see Monod–Cassidy (1980).
- (1771). *Astronomie*, 2nd edition, 2 vols. Paris.

Mason, Charles (1787). *Mayer's Tables improved by Mr. Charles Mason*, London: Commissioners of Longitude, 5s.6d.

Mayer, Tobias
- (1767). *Theoria Lunae juxta Systema Newtonianum*, London: Commissioners of Longitude. 2s.6d.
- (1770). *Tabulae Motuum Solis et Lunae/New and Concise Tables of the Motions of the Sun and Moon, by Tobias Mayer, to which is added The Method of Finding Longitude Improved by the Same Author.* London: Commissioners of Longitude. 5s.
- (1787), see Mason, Charles (1787).

Mudge, Thomas, Junior
- (1792a). *A Narrative of the Facts relating to some Time-keepers constructed by Mr. Thomas Mudge for the Discovery of the Longitude at Sea: together with Observations upon the Conduct of the Astronomer Royal respecting them.* London: for the author.
- (1792b). *A Reply to the Answer of the Rev. Dr. Maskelyne, Astronomer Royal, to a Narrative of Facts, etc., etc... To which is added a short Explanation of the most proper Methods of calculating a mean daily Rate, with Remarks on some Passages in Dr. M's Answer.* London: for the author.
- (1799). *A Description with Plates of the Time-keeper invented by the late Mr Thomas Mudge. To which is prefixed a Narrative by Thomas Mudge, his son...* London: for the author.

Nicholson's Journal (1806). Outline of the principal Inventions by which Timekeepers have been brought to the present degree of Perfection. Received from a Correspondent. *Jnl. of Nat.Phil., Chem., & Arts*, **14**, 273–302.

Ramsden, Jesse

- (1777). *Description of an Engine for dividing mathematical Instruments*. London: Commissioners of Longitude. 5s.
- (1779). *Description of an Engine for dividing Strait lines on Mathematical Instruments*. London: Commissioners of Longitude. 5s.

Rees, Abraham (1802-20). *Cyclopaedia*, Vol. 22 (1812), article 'Maskelyne'.

Robertson, John
- (1754). *Elements of Navigation*. 1st edition, 2 volumes.
- (1792). -, 2nd edition.

Royal Society, a member of (1763). *An Account of the Proceedings, in order to the Discovery of the Longitude at Sea*... London.

Russell, John (1797). *A Description of the Selenographia: an apparatus for exhibiting the phenomena of the Moon*... London: W. Faden.

Shepherd, A. (ed.) (1772). *Tables for correcting the Apparent distance of the Moon and a star from the effects of Refraction and Parallax*. London: Commissioners of Longitude. 21s.

Taylor, Michael (1780). *A Sexagesimal Table, exhibiting at Sight the Result of any Proportion where the terms do not exceed 60 minutes; and a Millesimal Table of Proportional Parts, with other useful Tables*. London: Commissioners of Longitude, 15s.
- (1792). *Table of Logarithms of all numbers, from 1 to 101000; and the Sines & Tangents to every second of the Quadrant*... with Preface and Precepts for the Explanation and use of the same by Nevil Maskelyne. London: Francis Wingrave. 3 gns. in sheets to subscribers, 4 gns. to non-subscribers.

Secondary published works

Allibone, T.E. (1976). *The Royal Society and its Dining Clubs*, London: Pergamon Press.

Arnold-Forster, Mary (1950). *Basset Down: An old country house*, London: Country Life.

Bence-Jones, Mark (1974). *Clive of India*. London: Constable.

Betts, Jonathan (ed.) (1984). *Principles and explanation of timekeepers by Harrison, Arnold and Earnshaw*. [Upton Hall, Newark-on-Trent]: British Horological Institute.

Cameron, H.C. (1952). *Sir Joseph Banks, KB, PRS: the Autocrat of the Philosophers*. London: the Batchworth press.

Cook, Andrew S. (1985). Alexander Dalrymple and John Arnold. *Vistas in Astronomy* (1985) **28**, 1/2, 189-95.

Cotter, Charles H. (1968). *A History of Nautical Astronomy*. London, Hollis & Carter.

Davies, R.D. (1985). A Commemoration of Maskelyne at Schiehallion. *Q.Jl.R.astr.Soc.* (1985) **26**, 289-94.

Dawson, Warren R. (ed.) (1958). *The Banks Letters*, London: British Museum.

Delambre, J.B. (1827). *Histoire de l'astronomie au XIIIe siècle*, Paris: Bachelier, 623-34.

Forbes, Eric G.
- (1965). Dr. Bradley's Astronomical Observations. *Q.Jl.R.astr.Soc.* (1965), **6**, 321-8.
- (1971a). Who discovered longitude at sea? *Sky & Telescope*, (January 1971), 4-6.
- (1971b). The Correspondence between Carl Friedrich Gauss and the Rev. Nevil Maskelyne (1802-5), *Annals of Science*, **27** (Sept. 1971), 213-37.
- (1974). *The Birth of Navigational Science*. Greenwich: National Maritime Museum, Monograph No. 10.
- (1975). *Greenwich Observatory, Volume 1: Origins and Early History*. London: Taylor & Francis.

Bibliography

Gould, Rupert T. (1923). *The Marine Chronometer, its History and Development*, London: Holland Press reprint 1960–78.

Howse, Derek, & Hutchinson, Beresford (1969). *The Clocks and Watches of Captain James Cook, 1769–1939*, London: Antiquarian Horological Society.

Howse, Derek (1975a). *Greenwich Observatory, Volume 3: The Building and Instruments*, London: Taylor & Francis.

– (1975b). *Francis Place and the early history of the Greenwich Observatory*. New York: Science History Publications.

– (1980). *Greenwich time and the discovery of the longitude*. Oxford, New York, Toronto, Melbourne: Oxford University Press.

Landes, David S. (1983). *Revolution in Time: Clocks and the Making of the Modern World*, Cambridge, Massachusetts: Harvard University Press.

Lane Hall, Mrs A.W. (1932). Nevil Maskelyne. *J. Brit.Astr.Ass.*, (Dec. 1932), 43, 2, 67–77.

Laurie, Philip (1967). The Board of Visitors of the Royal Observatory – 1710–1830. *Q.Jl.R.Astr.Soc.*, 7, 169–85.

Leadstone, G.S. (1974). Maskelyne's Schehallien experiment of 1774. *Physics Education* (1974), 9, 452–8.

Marguet, F. (1931). *Histoire Générale de la Navigation du XVe au XXe siècle*. Paris: Société d'Editions Géographiques, Maritimes et Coloniales.

Maskelyne, Thereza Story (1897). Nevil Maskelyne, D.D., F.R.S., Astronomer Royal. *Wiltshire Archaeological and Natural History Magazine* (June 1897), 29, 126–37.

May, W.E. (1976). How the Chronometer went to Sea. *Antiquarian Horology* (March 1976), 638–63.

Monod-Cassidy, Hélène (ed.) (1980). *Journal d'un Voyage en Angleterre 1763*. Oxford: The Voltaire foundation.

Quill, Humphrey (1966). *John Harrison, the Man who found Longitude*, London: John Baker.

Sadler, D.H. (1968). *Man is not lost*, London: HM Stationery Office.

Taylor, E.G.R. (1956). *The Haven-finding Art. A History of Navigation from Odysseus to Captain Cook*. London, Hollis & Carter. New edition 1971.

Weld, C.R. (1848). *A History of the Royal Society*, London, 2, 11–19.

Woolf, Harry (1959). *The Transits of Venus*, Princeton, N.J.: Princeton University Press.

Index

NM = Nevil Maskelyne: RO = Royal Observatory

Aberdeen observatory, 157
aberration of light, 10, 23, 223; defined, 237
Académie des Sciences, 46
account books, NM's, 113, 197, 249 (n.22); extracts from, 115, 128, 142
achromatic telescope, defined, 237
Act 12 Anne *c*.15, 11, 45, 76, 79, 126
Act 3 George III *c*.14, 45
Act 5 George III *c*.20, 79, 125
Act 13 George III *c*.77, 125
Act 14 George III *c*.64, 126, 170
Addington, Henry (1757-1844), 190, 191
Admiralty, 25, 27, 59, 110, 120
Advanced Room, 67, 72, 118
Adventure, discovery vessel, 123
Agria, Hungary, *see* Eger
Airy, George, FRS (1801-92), 116, 200; his transit circle, 67
Alarum Ridge observatory, St Helena, 31, 33, 37
altitude, defined, 237
Amelia Carolina, editor of *The Ladies' Magazine*, 10
American Academy of Arts and Sciences, 146, 161
American Independence, War of, 151
American Philosophical Society, 104
Amiens, Treaty of, 186, 195
Anderson, Prof. John, FRS (1726-96), 137
Andrews, Henry (1744-1820), 175, 201, 203
Anson, Adml. George, Lord, FRS (1697-1762), 14, 41
Armagh observatory, 157
Arnold, John, watchmaker (1736-99), 122, 127, 128, 171, 175, 189-92, 258 (n.17); chronometers by, 122, 123, 127, 152, 171, 179, 189; clocks by, 116, 119
Arnold, John Roger, watchmaker (-1843), 189-92
Arnold-Forster, Nigel, 145, 156, 164, 165, 194-5
Arrogant 74, 112

Ashurst, Capt., his barometer, 17
Assistants at RO, 56, 61, 64, 66, 73, 102, 118, 152, 154, 169, 200; their salaries, 61, 102, 118, 154, 201, 202
Assistant's library and calculating room, 64, 66, 73
Astronomer Royal, 53, 55; salary, 61, 121, 197, 201; *see also* Flamsteed, Halley, Bradley, Bliss, Maskelyne
Astronomer Royal for Ireland, 155
astronomical quadrant, 31
astronomical unit, 19; defined, 240
Astronomisches Jahrbuch, 149
Aswell, hatter, 116
Atkinson, Robert (W. Harrison's father-in-law), 74
Attraction of Mountains, 37, 129, 131, 134
Atwood, George, FRS (1746-1807), 174
Aubert, Alexander, FRS (1730-1805), 104, 186, 198, 209
Aurora frigate 32, 105
Ayscough, Mr, 5, 215
Ayscough, James, instmkr., 5
Ayscough, Rev. Samuel, 5

Baillie, Capt Thomas, RN, 81
Baltimore, Fredrick Calvert, 6th Earl (1731-71), 108, 109
Banks, Sir Joseph, Bart., PRS (1744-1820), 150, 160, 166, 171, 180, 181, 186, 187, 190-3, 205, 207; in the *Endeavour*, 110; on Attraction of Mountains committee, 132; becomes PRS, 147; in Mitre Club, 148, 150; relations with Herschel, 148; Paris/Greenwich connection, 151; Royal Society dissensions, 158-61, 255 (n.42); relations with Maskelyne, 160, 193, 201, 206, 210; the Mudge affair, 170-5; on Board of Longitude, 177, 190, 192-3; Arnold/Earnshaw affair, 190-3; appointed Associate of French Institute, 195; his *Protest...*, 190; and Maskelyne's

Index 269

library, 206; portrait, 166
Barbados trials, 1763–4, 40–52, 74, 75, 109; *see also* Observatories
Barham, Charles Middleton, 1st Baron (1726–1813), 192
Barrington, Hon. Daines, FRS (1727–1800), 131
Barrington, William Wildman, 2nd Viscount (1717–93), 75, 78
Basset Down, 100, 114, 163, 205
Batavia, 28
Bathe, Anne, 2
Bathe, William, 3
Bayly, William (1737–1810), Assistant at RO, 102, 118; transit of Venus in Norway, 104, 111; in the *Adventure*, 123; in the *Discovery*, 127; at Portsmouth Academy, 188
Bencoolen expedition, 22–4, 27, 216–7
Bennet, Mr, 17
Berge, Matthew, instrument maker, 169
Bernoulli, Jean, III (1744–1807), 104
Berthoud, Ferdinand, FRS, watchmaker (1727–1807), 46, 47, 80, 83
Bessel, Friedrich (1784–1846), 169
Betts, Rev. Joseph, 54, 56, 178
Bevis, John, MD, FRS (1709–76), 17, 45, 57, 74, 75, 110, 248 (n.10)
Billington, Mrs Elizabeth, soprano (1765–1818), 195
Birch, Thomas, DD, FRS (1705–66), 54, 245(n.18), 249 (n.20)
Bird, John (1709–76), 79, 128, 249(n.20); Board of Longitude award, 106; Greenwich instruments by, 64, 69, 70, 102, 198, 200, 223–4, 229–30; marine quadrants and sextants by, 15, 30, 49; other instruments by, 26, 49, 108, 112, 232, 250 (n.3)
Bishop, Robert, 45, 93, 94, 95
black-drop, 34, 105
Blagden, Sir Charles, MD, FRS (1748–1820), 149, 151, 152, 153, 160, 161
Blair, Rev. John, FRS (–1782), 54
Blake, Sir Francis, Bart., FRS (1708–80), 15
Bliss, Rev. Nathaniel, FRS (1700–64), 42, 44, 53–4, 58, 106; appointed Astronomer Royal, 42, 61, 219; at RO, 64, 70; on Board of Longitude, 44, 74, 82, 247 (n.12); dies, 53, 106, 221; his observations, 182
Bliss, Mrs, 106
Board of Longitude, 11, 14, 29, 41, 86, 96, 99, 105–6, 112, 120–1, 124, 157–8, 173, 177–80, 202; meetings, 47, 58, 59, 74, 75, 83, 84, 107, 171, 189, 190, 191, 202; observers, 48, 112, 123, 158, 178–9, 187, 188; publications, 83, 86, 106, 128, 157,
174, 178, 197; secretaries, 177
Board of Ordnance, 60, 61, 65, 70, 102, 199
Bode, Johann Elert, FRS (1747–1826), 149, 185
Booth, John (NM's grandfather), 3
Booth, Rev. Sir George, Bart. (NM's cousin) (–1799), 143, 163
Booth, Rt. Hon. George, Earl of Warrington (1675–1758), 142
Booth, Lady Letitia (née Rose), 142, 163
Booth, Rt. Hon. Nathaniel, Lord Delamere, 142
Borda, Jean Charles de (1733–99), 176
Boscovich, Roger Joseph, FRS (1711–87), 21, 23, 34, 104, 108, 130
Botany Bay, 158
Botolph Wharf below London Bridge, 28, 225
Bougainville, Louis-Antoine de, FRS (1729–1811), 176
Bouguer, Pierre, FRS (1698–1758), 108, 129, 131, 132
Bourne's house in Great College Street, Westminster, 5
Braden Farm, *see* Purton
Bradley, Rev. James, FRS (1693–1762), 10, 14, 15, 22, 23, 44, 54, 61, 64–5, 70, 82, 152, 155, 169, 199, 219, 223, 228–30; his instruments; 67, 68, 131, 198, 223, 228–30; his transit clock, 229; dies, 42; his observations, publication of, 43, 44, 55, 106, 121, 152, 180–2, 219–20
Bradley, John, 48, 50, 60, 74, 111, 121, 247 (n.7), 250 (n.3)
Bradley, Miss, 107
Bragge, MP, 173
Brest, blockade of, 15
Brewerton, robe maker, 116
Brice, Rev., 137
Bridgetown, Barbados, 49
Brilliant 36, 28
Brinkley, John, FRS (1763–1835), 155, 167, 178
Britannia merchantman, 52
British Catalogue of stars *see* Flamsteed
British Lying-in Hospital, 146, 205
British Mariner's Guide, *see* Maskelyne
British Museum, 106
Broughton, Capt. William Robert, RN (1762–1821), 179
Brühl, Johann Moritz, Count de, FRS (1736–1809), 80, 170–1, 174
Bryce, Mr, surveyor, 153
Bugge, Thomas, FRS, 119, 149
Bureau des Longitudes, 170, 176–7, 200
Bürg, Johann Tobias (1766–1834), 175, 200
Burke, Edmund (1729–97), 124
Burney, Dr. Charles jr., FRS (1757–1817), 200
Burrow, James, FRS (1701–82), 22, 23, 249 (n.20)

Burrow, Reuben, 118, 132, 134–7, 140–1
Byrne, Mary, miniaturist, 194

Calcutta, 34
callico, 100
Cambridge, Adolphus Frederick, Duke of (1774–1850), 198
Cambridge University, 6, 9, 54, 215
camera obscura, 5, 65, 118, 206, 215, 243 (n.9)
Campbell, Capt. John, RN, FRS (1720–90), 54, 249 (n.20); longitude trials, 14, 29; commands the *Dorsetshire*, 40, 48; Harrison calculations, 74, 78; tries Mudge chronometer, 171, 174
Camus, Charles-Etienne-Louis, FRS (1699–1768), 46, 47
Canton, John, FRS (1718–72), 54
Cape of Good Hope, 28, 34, 37, 41, 91, 108, 109, 158, 187, 216–7, 227
Carcass bomb 8, 127, 173
Caroché, Noël-Simon, instmkr., 176
Caroline, Princess of Wales (1768–1821), 198
Caroline of Anspach, Queen (1683–1737), 61
Carter, Richard (NM's cousin), 116, 205
Cassini family, 108
Cassini, Jean-Dominique, FRS (1748–1845), 152, 156
Cassini de Thury, César-François, FRS (1714–84), 151, 152, 176
Castries, maréchal de, 121
Catalani, Mme., singer, 208
Catharine Hall, Cambridge, 9
Cato, 188
Cavallo, Tiberius (1749–1809), 149, 174
Cavan, near Strabane, 110
Cavendish, Lord Charles, FRS, 54, 130, 247 (n.16), 249 (n.20)
Cavendish, Hon. Henry, FRS (1731–1810), 130, 131, 141, 151, 153, 157, 160, 195, 254 (n.18)
Centaur 74, 29
Centurion 60, 14
Ceres, minor planet, 184–6, 195
Chabert, Adml. Joseph-Bernard, marquis de, FRS (1724–1805), 170, 194, 210
Chambers, Sir William, FRS (1726–96), 161
Champlain, Lake, 161
Chandler, Dr, FRS, 249 (n.20)
Chapman, Benedict, assistant, 169
Chappe d'Auteroche, Jean-Baptiste (1728–69), 34
Charles II of England (1630–85), 11, 58, 62, 118
Charlotte, Queen (1744–1818), 198
Chatham brig 4, 178–9, 244 (n.5)
Chicheley, Sir Thomas (1618–99), 62
Chimborazo, mountain, 129
Chipping Barnet, 10, 11, 42, 43, 47

Choiseul-Amboise, Etienne François, duc de (1719–85), 80
Christ's Hospital, 157, 177
chromatic aberration, defined, 237
chronometers *see* Arnold, Berthoud, Earnshaw, Emery, Harrison, Kendall, Le Roy, Mudge
chronometer method for longitude, 14, 93, 126–7
chronometer, marine, defined, 237
chronometer rating, 256 (n.17)
Churchill, Charles, poet (1731–64), 5
circle, mural, 167, 200, 203
Circle Room, 65, 200, 203, 250 (n.15)
Clive, Edward, 2nd Lord Clive (later Earl of Powis) (1754–1839), 115
Clive, Margaret, Lady (née Maskelyne) (1735–1817), 2, 3, 6–8, 53, 100, 114, 117, 144–5, 163, 195, 205, 252 (n.1); portrait, 7
Clive, Robert, Lord (1725–74), 6–8, 57, 100, 101, 114, 115, 128, 243 (n.16), 252 (n.1)
clocks, *see* Arnold, Ellicott, Graham, Shelton
Cockburn, George, 76
comets, 17, 67, 72–3, 116, 146, 155, 158
compass, variation, 32
computers and comparers of Nautical Almanac, 85, 123, 175–6, 200; their salaries, 86, 157, 197, 201
Connaissance des Temps, 30, 41, 46, 87, 90, 120, 156, 176, 210
Constitution Hill, Barbados, observatory, 50
Cook, Capt. James, RN, FRS (1728–79); first voyage, 87, 116, 122, 157; second voyage, 87, 122–4, 126; third voyage, 87, 127, 157
Coombe, William, clkmkr., 127, 168
Copernican system, 23
Copland, Patrick (1749–1822), 137, 157
Copley medal, 139–40, 148
Cotterstock, 143, 163
Coutts, Thomas & Co., bankers, 170, 182
Covent Garden theatre, 208
Cowper, William, poet (1731–1800), 5
Crabtree, William (1610–44?), 18
Crane Court, London, 24, 161
Cricklade, 1, 2
Crocket, William, manservant, 128
Crosley, John, 155, 169, 175, 179, 187
Crown and Anchor tavern, 161
Cruizer sloop 8, 112
Cumberland schooner, 188
Cumming, Alexander, clkmkr., 247 (n.16)
Cust, Sir John, Bart., Speaker (1718–70), 75, 83

Daedelus storeship, 179
Dalby, Isaac (1744–1824), surveyor, 153
Dallaway, William, 106–7
Dalrymple, Alexander (1737–1808), 258

(n.17); his *A full answer...*, 193
Dance, Sir Nathaniel, HEIC (1748–1827), 188
Dartmouth College, New Hampshire, 161
Darwin, Erasmus, MD, FRS (1731–1802), 9
Dashwood, Sir Francis (1708–1781), 45
Davall, Peter, FRS, 226, 228
Davis, Capt Hesketh, 52
Davy, Sir Humphry, FRS (1778–1829), 203, 205
Dawes Point, Sydney, 158
Dawes, William, 158, 187
Dawes, William Rutter, FRS (1799–1868), 158
days: astronomical, civil, nautical, 245–6 (n.8)
de Luc, Jean André, FRS (1727–1817), 174
deck watch, 94; defined, 238
declination, celestial, 32, 64–6; defined, 238
declination, magnetic, *see* variation, magnetic
Delafaye, Charles, 3
Delambre, Jean-Baptiste Joseph, FRS (1749–1822), 176, 200 his *éloge* to Maskelyne, 5, 170, 210, 214
Delamere *see* Booth
Delisle, Joseph-Nicolas, FRS (1688–1768), 20, 21
Delmarva Peninsula, Maryland, 109
Demainbray, Stephen Charles Triboutet, (1710–82), 54
de Morgan, Augustus, FRS (1806–71), 140
density of the Earth, 131, 139–40
Deptford 60, 5
Devonshire, Duke of, 130
Dewar, Capt. James, HEIC, 38
Dickenson College, Pennsylvania, 161
Dickie, Andrew, watchmaker, 247 (n.16)
dip, magnetic, 32; defined, 238
dip circle or dipping needle, 32; defined, 238
dip of sea horizon, 92
Discovery discovery vessel 8, 127
Discovery sloop 10, 178
diurnal movement, 72; defined, 238
dividing engines, 127–8
Dixon, Jeremiah, FRS (1733–79), 26, 27–8, 37, 38, 108–9, 130, 216–7, 227
Dollond, John, FRS (1706–61), 70
Dollond, Peter (1730–1820), 30, 70, 71, 102, 104, 105, 118, 154, 227, 245 (n.21); micrometers by, 23, 33
Dorrington East Indiaman, 6
Dorsetshire 70, 40, 48
Down Farm, Purton, *see* Purton Down
Downman, John, ARA (1750–1824), 117
Drury Lane theatre, 197
Dublin, Trinity College, 167
Dunn, Samuel (1723–94), 94, 105
Dunsink Observatory, 166
Dunthorne, Richard (1711–75), 86, 91
Dutton, William, clkmkr., 174
Dymond, Joseph, 60, 100, 102, 110

Earnshaw, Thomas, watchmaker (1749–1829), 179, 188–93; his *Appeal to the Public*, 193
East India Company, 4, 21, 22, 31, 42, 77, 94, 188, 193, 218–9
eclipses of the Sun and Moon, 5, 67, 215
eclipses of Jupiter's satellites, 32; defined, 238; *see also* Jupiter's satellite observations
École Militaire, Paris, 176
Edwards, Mrs Mary, computer, 175
Egdon, Capt. S.R., 109, 252 (n.27)
Eger observatory, Hungary, 119
Egmont, John Perceval, 2nd Earl, FRS (1711–70), 53, 58, 75, 107
Elector Palatine, 119
Ellicott, John, FRS, clkmkr. (1706?–72), 225; clock by, 26, 37, 112; watch by, 49
Elliot, Sir Gilbert (1751–1814), 174
Ellis merchantman, 109, 252 (n.27)
Emerald 32, 111
Emery, Josiah, clkmkr., 175
Endeavour bark 6, 110, 122
Enjouée, French frigate, 152
ephemeris, defined, 238
equal altitude instrument, 31, 33, 37; defined, 238
equal altitude observations, 31
equation of time, 93; defined, 238
equatorial mounting, 32; defined, 238
equatorial sector, 67, 116, 154; defined, 241
Essex 70, 14
Etherington, Col., FRS, 161
Euler, Leonhard, FRS (1707–83), 78, 79
Evans, Rev. Lewis, FRS (1755–1827), 179
Evans, Thomas simpson (1777–1818), 73, 179
Everard, Edmund, 153
Ewing, John, 104, 119, 161

Ferdinand I of the Two Sicilies (1751–1825), 80, 185
Ferguson, James, FRS (1710–76), 73, 110
Ferminger, Thomas, assistant, 170, 200
Field, Governor Gregory, HEIC, 30, 259 (n.3)
figure of the Earth, 32; defined, 238
fireballs, *see* meteors
First Fleet to Australia, 158
First Point of Aries, 66; defined, 238
Flamsteed, Rev. John, FRS (1646–1719); becomes astronomer royal, 11, 58, 62–3; salary, 61; his instruments, 63, 222, 243 (n.9); dies, 63; his British Catalogue of stars, 18, 44, 63, 107; his books and papers, 43–4, 55, 107–8
Flamsteed House, 63, 64, 65, 66, 100, 155, 168, 198
Fleet, Mr, HEIC clerk, 225
Flinders, Capt. Matthew, RN (1774–1814), 187–8, 189
Floyer, Mrs Blanche, 114

Floyer, Miss, *see* Maskelyne, Mrs Edmund
Forbes, Prof. Eric Gray (1933–84), 77, 125
Forbes, Adml. Hon. John (1714–96), 76
force of gravity, 32, 33, 214–7
Fotheringham, Lt Patrick, RN, 48
Foundling Hospital, London, 44
Fowke, Eliza (*née* Walsh, NM's cousin) (1731–60), 6
Fox, Charles James (1749–1806), 173
Franklin, Benjamin (1706–90), 100, 104, 110, 131, 157
Frederick II, the Great, of Prussia (1712–86), 57, 104, 248 (n.10)
Friday Club, 161, 193–4
Frodsham, William, clkmkr., 82, 247 (n.16)
Furneaux, Capt. Tobias, RN (1735–81), 123

Garden Island observatory, Sydney, 187
Garnett, Joseph, assistant, 169
Gauss, Johann Karl Friedrich, FRS (1777–1855), 185, 186
Gentleman's Magazine, The, 17, 21, 56
George I of England (1660–1727), 11
George II of England (1682–1760), 11, 24
George III of England (1738–1820), 83, 131, 132, 151, 152, 167, 202, 222; grants for transits of Venus, 25, 110, 131; appoints NM as Astronomer Royal, 53, 58; RO regulations, 55–6; visits RO, 102, 198; petitions to, 118, 125, 173, 175; and Herschel, 148–9; and Ramsden, 169; his private observatory at Richmond, 124; his watch, 152; *see also* Royal warrants
George, Prince of Wales and Prince Regent (later George IV) (1762–1830), 198, 205
Georgium sidus, planet, 49; *see also* Uranus
Gibbon, Edward, historian (1737–94), 6
Gilpin, George, Assistant, 119, 154, 175, 177, 191–2
Glasse, Cambridge, tutor, 9
Glatton East Indiaman, 187
Glenie, Mr, 162
Globe Tavern, 161, 196
Gooch, William, 178–9, 210
Gooch, William the elder, 179
Gordon, Harry, Trinity College butler, 115
Gorgon 44, 158
Göttingen Royal Society, 120
Graham, Aaron, 174, 175
Graham, George, FRS, clkmkr. (1675–1751), 216, 222–3; clocks by, 64–5, 69, 73, 200, 222–3, 229–30; instruments by, 64, 67, 116, 167, 229, 230
Granby, Marquess of, 60
gravity, measurement, 32, 33, 37, 43, 216–7
Gray, Charles, 192
Great Barrier Reef, 188
great circle, defined, 239

Great Room, 63, 64, 65, 104, 116, 127, 230
Green, Charles (1735–71); Barbados trials, 48–52, 74, 221; Assistant at RO, 60, 104, 182; dies in the *Endeavour*, 110, 122, 123
Green, James, watchmaker, 247 (n.16)
Green and Blue, chronometers, *see* Mudge
Greenwich apparent time, 87; defined, 239
Greenwich Hospital, 81, 84
Greenwich mean time (GMT), 87; defined, 239
Greenwich Meridian, 59
Greenwich Observations, 102–3, 169, 181, 200, 222
Greenwich Park, 63, 149, 198–9
Grégoire, Citizen, 176
Gregor, MP, 173
Gregorian reflecting telescope, 31; defined, 239
Gregory, Olinthus, 160, 206
Grenville, Rt Hon. George (1712–70), 42, 53
Grignon, watch by, 49
Grove, Rev. Samuel, 10

Hadley, John, FRS (1682–1744), 96; *see also* quadrant, reflecting, *and* sextant
Haggis, Capt. Charles, HEIC, 28
Haley, Charles, watchmaker, 175
Halifax packet boat, 109
Halkhead, Mr, 163
Halley, Edmond, FRS (1656–1742); in St Helena, 30, 218; and transit of Venus, 18–29, 215–8; at RO, 11, 54, 63–4; salary, 61; founds Royal Society Club, 24; his mural quadrant, 64, 167, 222, 229; his clocks, 64, 73, 222, 229–30; his transit instrument, 223, 230; his observation books, 107
Halley's Mount, St Helena, 31, 33
Hamilton, James Archibald (1747–1815), 157
Hamilton, Sir William, FRS (1730–1803), 5
Hammerfest Island, 111
Hanmer, Rev. Thomas, 115
Harding, Carl Ludwig, FRS, 186
Hardy, William, clkmkr., 200
Harriot, Thomas (1560–1621), 256 (n.15)
Harrison, John, clkmkr. (1693–1776), 96, 172, 210, 220ff; early work, 14; Jamaica trial, 40–1, 220; affairs 1763, 44–6; Barbados trials, 48–51, 220; affairs 1765–7, 75–84, 221–2; affairs 1772–4, 124–6; dies, 126; chronometers by: H1, 82; H e3, 14, 40, 82; H4, 14, 40–1, 45–8, 50–1, 74–5, 79–84, 122, 127, 154; H5, 123, 124; gridiron pendulum, 223; his *An Account of the Proceedings . . .*, 45, 77; his *Remarks on a Pamphlet . . .*, 84; his *The Principles of Mr. Harrison's Time-keeper*, 87
Harrison, William, FRS (c.1730–1815), 44, 77;

Jamaica trials, 40-1; Barbados trials, 48-51; his manuscript journal, 50
Harvard College, 34, 161
Hassell, Richard, FRS (-1770), 15, 16, 17, 42
Hastings, Warren (1732-1818), 5
Hawke, Adml. Edward, Baron (1705-81), 15
Heberden, William, MD, FRS (1701-1801), 54, 249 (n.20)
Hellins, John, assistant, 119, 138, 154
Hengest, Lt. Richard, RN, 179
Henn, Martha *see* Rose
Henry Dundas East Indiaman, 154
Herschel, William, FRS (1738-1822), 146-9, 154, 155, 157, 161, 164, 166-7, 186, 209
Herschel, Caroline Lucretia (1750-1848), 148, 155
Hirst, Rev. William, FRS (-1770), 35, 105
Hitchins, Rev. Malachy, computer (1741-1809), 104, 175, 197, 201
Hoares, Messrs., bankers, 182
Hodgson, Friday Club member, 161
Holmes, John, watchmaker, 175
Hooke, Robert, FRS (1635-1703), 230
Hornsby, Rev. Thomas, DD, FRS (1733-1810), 76, 107, 122, 152, 171, 175, 180-1, 190, 203, 220
Horrocks, Jeremiah (1617?-41), 18
Horsley, John, 104
Horsley, Rev. Samuel, FRS (1733-1806), 131, 159-60, 174, 209, 255 (n.42)
Houblon, Sir James (-1700), 2
Hounslow Heath, 152
Howard, Mrs Katharine (NM's great-aunt), 5
Howe, Adml Richard, Earl (1726-99), 54, 82
Howells, William, watchmaker, 175
Huddesford, Rev. G., 142
Hudson's Bay Company, 110
Hunt, master RN, 112
Hunter, Capt. John, RN (1738-1821), 158
Hurd, Capt. Thomas, RN (1757?-1823), 177
Hutchinson, Governor Charles, HEIC, 30, 34, 38, 218
Hutton, Prof. Charles, FRS (1737-1823), 141, 159-60, 161, 175, 205, 209, 234, 253 (n.10)

Ibbetson, John, 81, 177
inclination, magnetic, *see* dip
income tax, 195
Inman, Rev. James (1776-1859), 187-8
Institut National, Paris, 195, 210
internal contact, 31; defined, 239
Investigator survey ship 22, 187-8, 189
Irwin, Christopher, his marine chair, 47, 48-9, 74, 221

Jamaica, 40, 48
Jamestown, St Helena, 29, 30, 35, 37; James's Fort, 33, 37; Sisters' Walk, 33
Jeaurat, Edme-Sébastien, 121, 156
Juno, minor planet, 186
Jupiter's satellite observations, 32, 47, 49, 67, 72-3, 221

Keech, Joseph, computer, 85, 87
Kelsall, Alice (*née* Maskelyne) (NM's aunt) (1699-), 2
Kelsall, Jane (Jenny) (NM's cousin) (1740-1824), *see* Strachey)
Kendall, Larcum, 79, 80, 81, 84, 122, 168; chronometers by: *K1*, 84, 122, 123, 126, 127, 158; *K2*, 127; *K3*, 127, 179, 187
Kensington Gore, 3
Keroualle, Louise de, Duchess of Portsmouth (1649-1734), 62
King, Lieut James, RN (1750-84), 127
Kinloch Rannoch, 134-7
Kinnebrook, David the younger (-1802), 169
Kippis, Andrew, DD (1725-95), 255 (n.42)
Knight, attorney in Barbados, 50
Knowles, Adml Charles (1700?-77), 40, 76

Lacaille, Nicolas-Louis de, FRS (1713-62), 41, 90, 108, 152, 210; his *Fundamenta Astronomiae*, 44
La Condamine, Charles-Marie de, FRS (1701-74), 47, 108, 129
Lady's Magazine, The, 9
Lagrange, Joseph-Louis Marie, FRS (1736-1813), 176
Lake Champlain, 161
Lalande, Joseph-Jérôme Lefrançais de, FRS (1732-1807), 5, 83, 108, 148, 156, 176, 210; editor of *Connaissance des Temps*, 41, 87, 91, 120-1; visits England, 46, 47, 156; his *Astronomie*, 175, 176
Laplace, Pierre-Simon, marquis de, FRS (1749-1827), 108, 175, 176
Laplace/Bürg tables, 175, 200
Large, Mr & Mrs Purton Stoke, 163
Larkins, William (-1798), 170
Latham, Jane (Jenny) (*née* Kelsall), *see* Strachey
Latham, Capt. Thomas, RN (-1762), 8
Lax, William, FRS (1761-1836), 178, 190
Le Gentil de la Galaisière, Guillaume-Joseph-Hyacinthe-Jean-Baptiste (1725-92), 21, 34
Le Grand French frigate, 27
Le Maire, Madame, 143
Le Roy, Pierre (1717-85), 83
Lee, Stephen, 186
Legendre, Adrien-Marie, FRS (1752-1833), 154, 156
Leigh and Sotheby, auctioneers, 206

L'Emery, 121
lemon juice, 25
Lemonnier, Pierre-Charles (1715-99), 83
Lichfield, Henry Lee, 3rd Earl (1718-72), 54
Limb, 33; defined, 239
Lind, James, MD, FRS (1736-1812), 131, 149-50
Lindley, Joseph, assistant, 152, 154, 175
Lindsay, Capt. Sir John, RN, MP (1737-88), 50-1, 83
Linois, Adml, French navy, 188
liquor, 25
Lizard Point, 111
Lockman, Rev. John, DD, FRS, 149
logarithm tables, 91, 178
logarithms, proportional, 92
London, 30
London Bridge, 184
Long, Prof. Roger, FRS (1680-1770), 55, 81-2
Longitude Acts, *see* Act(s)
longitude at sea, 11; by chronometer, 13-14, 93; by lunar observations, 13-14, 30, 38, 42, 48, 77, 92-3
Longitudes, Bureau des, see Bureau
Louis XIV of France (1638-1715), 62
Lowndes Professor at Cambridge, 178
Lucasian Professor at Cambridge, 76, 178
Ludlam, Rev. William (1717-88), 79, 80, 81
lunar parallax, 37, 91, 94
lunar-distance tables, 42, 87, 89-93, 120-1, 210
Lutwidge, Capt. Skeffington, RN, 127
Lying-in Hospital, *see* British
Lyons, Israel the younger (1739-75), 86, 91, 127

Macclesfield, George Parker, 2nd Earl, PRS (1697-1764), 22-3, 26, 42, 44, 53, 219, 247 (n.12)
Mackenzie, James Stuart, 137, 232, 247 (n.12)
Madras, 6, 244 (n.5)
Magnamine 74, 48
magnetic observations, 32
Maire, Christopher (1697-1767), 108, 130
Manley, Capt. Robert, 52
Mannheim observatory, 119
Mapson, John, computer, 86, 111
Marguet, F., capt. de vaisseau, 91
Marine Society, 146
Mascall, Deborah, 73, 170
Maseres, Francis, Baron of the Exchequer, FRS (1731-1824), 161, 193, 209
Maskelyne family, general, 1-8, 99-102, 113-5, 142-6, 206
Maskelyne, Alice (NM's aunt) (1699-   ), *see* Kelsall
Maskelyne, Ann (James Houblon's wife) (1730-1813), 143, 205
Maskelyne, Ann (James Houblon's daughter) (1761-   ), 143, 205
Maskelyne, Anne (NM's aunt) (1694-1767?), 2, 6
Maskelyne, Anne (*née* Bathe) (NM's grandmother) (-1706), 2
Maskelyne, Catharine (*née* Greenly, then Muscott) (-1786) (Edmund's 2nd wife), 112, 114, 163
Maskelyne, Edmund (NM's father) (1698-1744), 3, 243 (n.8)
Maskelyne, Edmund (NM's brother) (1728-75), 3-6, 48, 53, 100, 112-4, 162, 243 (n.8)
Maskelyne, Mrs Edmund (*née* Floyer) (Edmund's 1st wife) (-1762), 100
Maskelyne, Elizabeth (*née* Booth) (NM's mother) (-1748-9), 3-5, 243 (n.8)
Maskelyne, Elizabeth (NM's aunt) (1697-1734), *see* Walsh
Maskelyne, James Houblon (NM's uncle) (1701-76), 2, 99, 114, 143, 205
Maskelyne, Jane (NM's aunt) (1693-1766?), 2, 99
Maskelyne, Jane (James Houblon's daughter) (1764-1805), *see* Toomer
Maskelyne, Jasper (NM's cousin), 164
Maskelyne, John Nevil, illusionist (1839-1917), 164
Maskelyne, Margaret (NM's sister) (1735-1817), *see* Clive
Maskelyne, Margaret (NM's daughter) (1785-1858), 73, 117, 144, 163-4, 193-5, 203, 205-6, 212, 236, 259 (n.7); born, 143; marries, 206; portraits, 145, 165
Maskelyne, Nevil (NM's great-great-grandfather) (1611-79), 235
Maskelyne, Nevill (NM's grandfather) (1661-1711), 1
Maskelyne, Nevil (NM's uncle) (1692-1774), 2, 6, 10, 114
Maskelyne, Rev. Nevil, DD, FRS (1732-1811); born, 3, 243 (n.6); christened, 3; school days, 3-6, 214-5; to Cambridge, 9-10, 215; BA, 9, 215; MA, 10, 15; Fellow of Trinity, 10, 215; takes Holy Orders, 10; curate in Chipping Barnet, 10, 16-17, 42-3, 47; chosen for transit of Venus, 24, 215; St Helena, 28-38, 216-8; expenses there, 25, 225-8; German soldier there, 38, 217-8; addresses HEIC Court of Directors, 42, 218; addresses Royal Society, 43, 219; Barbados, 48-53, 220-1; appointed Astronomer Royal, 58-60, 219; took up residence, 60; BD, 102; DD, 115-6; awarded Copley medal, 139-40; marries, 142; dies, 203; buried, 205; his ailments, 162, 198; his autobiographical

notes, 1, 214–24; his crest and arms, 128; his friends, 209; his handwriting, 111, 204–5; his holidays, 4, 128, 163, 182, 195; his library, 206, 243 (n.1); his observing suit, 100–1; his salary, 61, 121, 195, 201; his will, 203, 205; foreign honours, 120, 195; memorial tablets, 235–6; portraits, Frontispiece, 117, 144, 164, 194, 196; his *A Plan for observing meteors called fireballs*, 151; his *An Account of the going...*, 84, 87; his *An answer to a pamphlet...*, 174; his *British Mariner's Guide*, 42, 77, 90, 94, 218; his *Arguments for giving a reward...*, 191, 193; *see also* Board of Longitude, Royal Observatory, Royal Society
Maskelyne, Nevil (NM's cousin), 164
Maskelyne, Sarah (NM's aunt) (1695–1767?), 2, 99
Maskelyne, Sophia (*née* Rose) (NM's wife) (1752–1821), 142–3, 162, 205–6, 236; portraits, 145, 194, 197
Maskelyne, Rev. William (NM's brother), 3–4, 5, 9, 24, 99–100, 113–4, 243 (n.8)
Maskelyne, Wynn (NM's uncle) (1704–38), 2
Maskelyne Islands, New Hebrides, 180; *see also* Mount Maskelyne
Maskelyne Point and islands, Portland Sound, Oregon, 179, 257 (n.36)
Maskelynge, Robert, 1
Maskelynge, William, of Purton, 1
Mason, Charles (1728–86); and 1761 transit of Venus, 24, 26, 27–8, 216–7, 227; and Mason–Dixon line, 108–9, 130; and 1769 transit of Venus, 110–1; and attraction of mountains, 131–2, 141; his *Mayer's lunar tables improved...*, 120, 157, 175; dies, 157–8
Mason, Mrs, 157
Mason–Dixon Line, 108–9, 141, 232
Masters of the Royal Navy, 94, 121
Matthews, William, wtchmkr., 79
Maty, Paul Henry, MD, FRS (1745–87), 159–60, 255 (n.42)
Mauduit, Israel, FRS (1707–87), 249 (n.20)
Maupertuis, Pierre Louis Moreau de (1698–59), 108
Mauritius, 188
Mayer, Johann Tobias (1723–62), 211; his first manuscript lunar and solar tables, 14, 29, 38, 42, 218–9; his last manuscript tables, 47, 49, 77, 78, 79, 86, 198, 222; dies, 41; receives longitude award, 77, 79, 125; his lunar tables improved by Mason, 120, 157, 175; his repeating circle, 14–15; his *Theoria Lunae Juxta...*, 86
Méchain, Pierre-François-André, FRS (1744–1804), 121, 147, 154, 156, 176, 186

Melville, Henry Dundas, 1st Viscount (1742–1811), 191–2
Memorandum books, NM's, 113; extracts from, 161–2, 182–3, 198, 207–8
memorial tablets, 235–6
Mendoza y Rios, Joseph, FRS, 194
Menzies, Sir Robert, 157
Menzies, William, surveyor, 135–6, 233
Menzies, Commissioner, 157
Mercury, planet, 18
meridian, defined, 239
meridian altitude, 908
Merlin sloop, 19, 41
Messier, Charles, FRS (1730–1817), 147
meteor, 149–51
Michell, John, BD, FRS (1724–93), 21, 43, 45, 79, 141, 244 (n.5), 247 (n.16)
micrometer, astronomical, 31, 33; defined, 239
Milford 28, 48
Milman Street, London, 143
Milner, Isaac, FRS (1750–1820), 178, 190
minor planets, 184–7
Minto, 1st earl, *see* Elliot
Mitre Club and tavern, *see* Royal Society Club
Molyneux, Samuel (1689–1728), 223
Monatliche Correspondenz..., 186
Montague House, Greenwich Park, 198
Moon globe and maps, *see* Russell
Moon's age, 90; defined 239
Moon's diurnal parallax in right ascension, 32, 37, 52, 221
Moore, Sir Jonas, FRS (1617–79), 61
Morgan, member of the Friday Club, 161
Morris, Gael, 30, 60
Morton, Charles, MD, FRS (1716–99), 27, 54, 249 (n.20)
Morton, James Douglas, 14th Earl of, PRS (1702–68), 53, 54, 76, 220, 221, 247 (n.16), 249 (n.20)
Moss, Rev. Charles, FRS, 22, 54
Mount Maskelyne, 180
Mount & Page, publishers, 86
Mountaine, William, FRS (–1779), 15
movable quadrant, *see* quadrant
Mudge, Thomas the elder (1717–90), 79, 80, 83, 126, 170–5, 247 (n.16); his chronometers, 127, 170–1, 189; dies, 175
Mudge, Thomas the younger (1760–1843), 172–5; his *A narrative of the facts...*, 172, 174; his *A reply to an answer...*, 174
mural circle, 198–9, 200, 203
mural quadrant, *see* quadrant
Murdoch, Rev. Patrick, FRs (–1774), 54
Muscott, Catharine, *see* Maskelyne, Catharine
Muscott (son), 114

Nairne, Edward, FRS (1726–1806), 105

Nautical Almanac, 42, 96, 171; foundation of, 59, 85, 222; description of, 85–96; publishing history, 93, 99, 103, 120–1, 157, 175–6, 200, 203, 222
Navy Board, 94, 121
Nelson, Vice Admiral, Viscount (1758–1805), 5, 208
Nevilles of Abergavenny, 2
Newcastle, Thomas Pelham Holles, FRS, Duke of (1693–1768), 3, 4, 24, 243 (n.8)
Newcastle House in Lincoln's Inn, 3
Newcastle Place, London, 194
New Elizabeth merchantman, 51
Newfoundland, 34, 171, 173
New Hebrides, 180
New Observatory, Greenwich, 64, 65, 66, 200
Newton, Sir Isaac, PRS (1642–1727), 43, 57, 129, 134, 139, 141, 159
Newtonian reflecting telescope, defined, 239
Nicholson's Journal, 193
Niebuhr, Carsten, 41
Nivernais, Louis-Jules, duc de, FRS, 46
Nootka incident, 178
North, Frederick, 8th Lord North and Earl of Guilford (1732–92), 119, 121, 124–5, 180, 220
North Runcton, 146, 163
Northumberland East Indiaman, 29
Norwood, Capt. Joseph, RN, 48
Nourse, John, bookseller, 86
nutation of the Earth's axis, 11; defined, 239

Oahu, 179
Oakly Park, near Ludlow, 114, 128
object glass, 35; defined, 239
objective, telescopic, defined, 239
Observatories: Aberdeen, 157; Armagh, 157; Barbados, Constitution Hill, 50, 75, Willoughby Fort, 50; Dunsink, 167; Eger, 119; Greenwich, *see* Royal Observatory; Mannheim, 119; Oxford (Radcliffe), 119; Palermo, 156, 184–5; Paris, 62, 151, 152, 176; Philadelphia, 104, 119; Portsmouth, Naval Academy *see* Royal Naval Academy; Richmond (Kew), 124; St Helena, Alarum Ridge, 31, 33, 37; Jamestown, 33, 35; Sandy Bay, 35, 38; Seeberg, 171; Sydney, NSW, Dawes Point, 158; Garden Island, 187
observing policy, NM's, 71–2
observing suit, NM's, 100–1
occultations, 32, 67, 72–3; defined, 239
octant, *see* quadrant, reflecting
Olbers, Wilhelm, FRS (1758–1840), 185–6
Ordnance Survey, 137, 151
Oriani, Barnaba, 185
Osborn, Adml Henry (1698?–1771), 76
Owen, William, RA, (1769–1825), 165
Oxford East Indiaman, 34

Oxford: Press, 180: Radcliffe observatory, 119; university of, 54

Palermo observatory, 156, 184–5
Palermo circle, 167
Pallas, minor planet, 186
parallax, 91; defined, 240; lunar, 32, 37, 52, 91; of Sirius, 22, 32, 34–5; solar, 20, 91; stellar, 23
Paramore pink, 6, 61
Paris, Peace of, 41
Paris–Greenwich connection, 64, 151–4
Paris Observatory, 62, 151, 152, 176
Parker, Sir Harry, Bt., 177
Parliament, and Harrison, 44–6, 78–9, 82–3, 124–6; and Mudge, 173–5; and Earnshaw/Arnold, 193; select committees, 174–5
Pate Rose, John, *see* Rose
Peach, Rev. Samuel (–1769), 107, 121
Pemberton, Henry, MD, FRS (1694–1771), 54
Pembroke Hall, Cambridge, 9
Penn, Thomas (1702–75), 108, 109
Penny post, 182
Penruddock, Colonel, 2
Perceval, Spencer (1762–1812), 201–2
Philanthropic Society, 146
Phillip, Capt. Arthur, RN (1738–1814), 158
Phipps, Capt. Constantine John, RN, 2nd Baron Mulgrave (1744–92), 127, 173
Piazzi, Giuseppe, FRS (1746–1826), 156–7, 167, 184–7
Picard, Jean (1620–82), 108
Pigott, Matthew, 245 (n.18)
Pigott, Nathaniel, FRS (–1804), 26, 149
Pingré, Alexandre-Gui (1711–96), 34
Pitt, William (1759–1806), 173–4
Pitts, John, 179
Playfair, John, FRS (1748–1819), 137, 141
Plumian Professor at Cambridge, 76, 177
Pocock, Adml Sir George (1706–92), 76
Point Maskelyne, 158
polar distance, defined, 240
Polwarth, Lord, FRS, 137
Pond, John, FRS (1767–1836), 199–200, 202, 205
Pondicherry, 21, 30, 34
Ponds farm, *see* Purton Stoke
Poor Orphans of Clergy, 146
Poore, Edward FRS, 159
Porpoise storeship 10, 188
Port Mahon 24, 29
Portland, William Cavendish, 3rd Duke of (1738–1809), 181
Portland Sound, 179
Portsmouth, Duchess of, *see* Kerouaille
Portsmouth, *see* Royal Naval Academy
Postage book, NM's, 155–6
Price, Mrs (*née* Halley), 107

Index

Prince Henry East Indiaman, 26, 28-30
Princess Louisa 60, 48-9, 91
Princess of Wales, *see* Caroline
Princeton university, 161
Pringle, Sir John, PRS (1707-82), 123, 139
prismatic micrometer, 120
proportional logarithms, 91
Providence sloop 12, 169, 179-80, 187
Purling, Matthew, HEIC, 33
Purton, 1; Church of St Mary, 2, 205, 206, 235; Sunday school, 146; Braden farm, 163; West Marsh, 2
Purton Down, 2, 48, 99, 114, 163, 205
Purton Stoke (Ponds farm), 3, 4, 100, 113, 114, 163, 206

Quadrant Room, 64-5, 229
quadrant, astronomical, 31; defined, 240
quadrant, movable, 65, 118, 167; defined, 240
quadrant, mural, 64, 68, 198, 200; defined, 240
quadrant, Hadley's reflecting, 13, 30, 42, 94, 120, 216; defined, 240
Quiberon Bay, Battle of, 15
Quill, Col. Humphrey, RM (-1987), 51, 125-6, 246

Racehorse bomb 8, 127
Radcliffe observatory, Oxford, 119
Ramsay, Prof., 137
Ramsden, Jesse, FRS (1735-1800), 167-8, 174; dines at RO, 157; dividing engines by, 127-8; instruments by, 147, 151-2, 153, 156, 167-8, 185, 232
Ramsgate, 195, 203, 204
Raper, Matthew, FRS, 54, 132
Rees, Abraham, DD, FRS (1743-1825), his *Cyclopaedia*, 214
reflecting circle, 14
reflecting telescope, 31, 104-5
refraction, atmospheric, 15, 91; defined, 240
Regulations for the Royal Observatory, 55-6, 60, 118
Reid, Prof. Thomas, FRS, 137
Rennie, John the elder (1761-1821), 184
Requisite tables, *see Tables Requisite...*
Resolution discovery vessel, 123, 127
Richardson & Clark, printers, 86
Richmond, Charles Lennox, 3rd Duke of, FRS (1735-1806), 50, 155
Rider, MP, 173
Rigaud, Stephen Peter, FRS (1774-1839), 203, 256 (n.15)
right ascension, 66; defined, 240
Rippon 29, 60
Robbins, Reuben, 85, 87
Roberts, Capt. Henry, RN, 178
Robertson, Rev. Abram, FRS (1751-1826), 15, 177, 181, 182, 190-1, 203, 208, 250 (notes 6 & 7)

Robertson, Duncan (of Rannoch), 137-8
Robertson, John, FRS (1712-76), 15, 247 (n.7); his *Elements of Navigation*, 94
Robinson, John, 124
Robinson, Charlotte and Harriet (NM's nieces), 205
Robinson (of Perth), 149
Robison, John (1739-1805), 40, 48, 60
Rockingham, Charles Watson Wentworth, Marquess of, FRS (1730-82), 5
Rodney, Adml George Brydges Rodney, 1st Baron (1719-92), 81
Rodriguez I., 34
Rolla, 188
roof shutter openings, 118
Rose, John Pate, 143
Rose, Letitia, *see* Booth
Rose, Martha Henn (NM's mother-in-law) (1723-83), 143, 254 (n.3)
Rose, Sophia, *see* Maskelyne
Ross, Rev. John D., DD, FRS, 100
Rous, Thomas, HEIC, 42
Rowley, Adml Sir William (1690?-1768), 75
Roy, Gen. William, FRS (1726-90), 64, 123, 137, 151-4, 156
Royal Astronomical Society, 136
Royal George 15, 100
Royal Greenwich Observatory, 1
Royal Military Academy, Woolwich, 141, 160
Royal Museum of Scotland, 232
Royal Naval Academy at Portsmouth, 43, 74, 94, 121, 188, 247 (n.7)
Royal Observatory, Greenwich, 18, 62-73, 81, 102-5, 116-20, 167-70, 198-201; instruments, 63-73, 102-5, 116-8, 154, 167, 198-200, 222-4, 229-31, 252 (n.9); observing policy, 71-2; regulations, 55-6, 60, 118; *see also* Advanced Room, Assistants, Assistant's Room, Circle Room, Flamsteed House, Great Room, New Observatory, Quadrant Room, Summer Houses, Transit Room, Visitations, Visitors
Royal Society, 136, 144, 150, 193, 203; and the 1761 transit of Venus, 22-8; and the Mason-Dixon line, 108-9; and the 1769 transit of Venus, 109-12; and Schiehallion, 130-2, 139-40; at Crane Court, 24, 161; at Somerset House, 161; and Bradley's observations, 106; dissensions in, 158-61; *see also* Visitors
Royal Society Club, 24, 43, 47, 100-2, 123, 130, 148, 150, 161, 179, 193, 203, 243 (n.12)
Royal warrants, 24, 58-9, 60, 121, 202
running fix, 93; defined, 241
Russell, John, RA (1745-1806), 164-7; his moon maps and globe (Selenographia), 164-7, 196; portraits by, Frontispiece, 166, 196, 197

Index

Sacheverell, Jane (*née* Secker), 100, 113–4, 145–6, 205, 252 (n.1)
Sacheverell, William, 113
St Andrew, Holborn, 142
St Helena, 22–4, 27–39, 44, 129, 215–8; expenses at, 226–8; instruments, 23–4, 31–2, 36, 109, 225–7, 245 (n.21)
St Helens Roads, 28, 48
St James's Chronicle, 140
St James's Fort, *see* James's Fort
St John, Henry, 1st Viscount Bolingbroke (1678–1751), 2
St Martin-in-the-Fields, Westminster, 3
St Petersburg Academy, 120
St Pierre, Sieur de, 62
St Vincent, Admiral John Jervis, 1st Earl (1735–1823), 190
Saintsbury's school in Cirencester, 6
Salaries: RO Assistants, 61, 102, 118, 154, 201, 202, 210; Astronomer Royal, 61, 121, 195, 201; computers, 86, 157, 176, 195–7, 201, 210
Sam, manservant, 194
Sandby, Paul, RA (1725–1809), 149, 150
Sandby, Thomas, RA (1721–98), 149, 150
Sandwich, John Montagu, 4th Earl of, FRS (1718–92), 53, 58, 60, 120, 122
Sandy Bay, St Helena, 35, 38
Savilian Professors at Oxford, 54, 58, 61, 177, 202
Saxe Gotha, Duke of, 171
Schiehallion, mountain, 37, 119, 132–8; instruments and equipment, 36, 232–4
Schwep, Mr, soda water maker, 207
Scott, George Lewis (170-8–80), 45, 54, 247 (n.16)
scurvy, 26
Seaford 22, 112
Seahorse 24, 25–8
Secker, Jane *see* Sacheverell, Jane
sector, *see* equatorial *and* zenith sectors
Seeberg observatory, 171
Selenographia, *see* Russell
semi-diameter, defined, 241
Seven Years War, 15, 21, 41
sextant, Hadley's reflecting, 15, 49
Sextant and Quadrant Houses, 16, 63, 66
Shelburne, William Petty, 2nd Earl (1737–1805), 108
Shelton, John, clkmkr., 102, 249 (n.20); clocks by, 26, 31, 32–3, 37–8, 49–50, 65, 109, 112, 134, 225, 232, 245 (n.21)
Shepherd, Anthony, BD, FRS (1721–96), 53–5, 58, 76, 78, 120, 162, 177
Sheridan, Richard Brinsley, dramatist (1751–1816), his *The Duenna*, 195
sherry, 162
Short, James, FRS (1710–68), 45, 46, 57, 74–5, 83, 110, 148, 247 (n.16); telescopes by,

26, 65, 102, 104, 112, 223, 230, 245 (n.21)
Shrawardine, 115, 146
Shuckburgh-Evelyn, Sir George Augustus William, FRS (1751–1804), 173, 209
Simpson, Thomas, FRS (1710–61), 15
Sinley, Mr, watercolourist, 156
Sirius, annual parallax of, 22, 32, 216
Sirius 22, 158
Sisson, Jeremiah, 48, 106, 119, 156; instruments by, 23, 32, 35, 36, 116, 131, 147, 156, 225–7, 232
Sisson, Jonathan, (1690–1749), 222
Sisters' Walk, *see* Jamestown
Smith, John, DD, 178
Smith, John, MD (–1797), 178
Smith, Robert, DD (1689–1768), 55, 115, 249 (n.6)
Smithson, James Louis Macie (1765–1829), 203
Smithsonian Institution, 203
Smyth, Adml William H., FRS (1788–1865), 26, 245 (n.15)
Society of Naval Architecture, 146
Solander, Daniel Carl (1733–82), 110, 123
solar parallax, defined, 240
Somerset House, 161
Speaker of the House of Commons, 75, 83, 120, 177
Spencer, Earl, 181
star catalogues, *see* Flamsteed
stellar parallax, defined, 240
Stephens, Philip (1725–1809), 76, 81
Stewart, Provost John, 233
Story, Anthony Mervyn, FRS (later Story-Maskelyne) (1791–1879), 206
Strachey, Jane (*née* Kelsall) (NM's cousin), 8
Stradivarius, Antonio, 138
Styles, Dr, of Yale, 161
Summer Houses, 65, 105, 116
Supply armed tender, 158
Swindon, 1
Sydney, NSW, 158, 187
Sykes, James, ship's agent, 179

Tables Requisite. . . , 85, 86, 91, 94, 157, 197–8
Tahiti, 110
Talmash, Mrs, landlady, 194
Tartar 28, 50, 83
Taylor, Prof. Eva G.R., her *The Haven-finding Art*, 96
Taylor, Henry, 143
Taylor, John Michael (Michael's son), 178
Taylor, Michael, computer, 171, 178, 210
Taylor, Thomas (Michael's father), 178
Taylor, Thomas, Assistant, 200
telescope, achromatic refracting, defined, 237, 241
telescope, Cassegrain reflecting, defined, 237, 241

telescope, Gregorian reflecting, 31; defined, 239, 241
telescope, Newtonian reflecting, defined, 239, 241
telescope, triple achromatic, 70-1, 104, 118
Telford, Thomas (1757-1834), 184
Terpsichore 24, 38-9
Tew, Mrs Elizabeth, 107
Theed, wig maker, 116, 194
Theed the younger, miniaturist, 194
Three Tuns Tavern, 9
tidal observations, 32, 37-8, 217, 246 (n.29)
Tilbury Fort, 63
Titius-Bode Law, 185; defined, 241
Tobolsk, 34
Toomer, Joseph, 143, 205
Toomer, Jane (*née* Maskelyne), 143, 206
Tornea, 34
Tothill Street, Westminster, 3
Tower of London, 63
Townshend, Charles, Viscount, 2
Trafalgar, Battle of, 207
transit, defined, 241
transit clock, defined, 241
transit instrument, 32, 65; defined, 241
Transit of Mercury, 18, 215
Transit of Venus (1761), 18-34, 215-8, 225-30
Transit of Venus (1769), 35, 104-5, 110-2
Transit Room, 64-9, 127, 189, 229
Treaty of Amiens, 186, 195
Treaty of Versailles, 151
trials, *see* Harrison, Mudge
Trinity College, Cambridge, 3, 9-10, 115, 146, 215
Trinity College, Dublin, 119, 139, 167
Troughton, Edward (1753-1835), 154, 167, 173, 187, 199 200, 203
Tuscany, Grand Duke of, 168
Tyrrell, Adml Richard (1717-66), 48-9; his pocket watch, 49

Uranus, 149, 185

Van der Puyl, Louis-François-Gérard, painter, 117, 144, 145, 187
Vancouver, Capt. George, RN (1758-98), 87, 178-9
Vansittart, Henry (1732-70), 105
variation, magnetic, defined, 241
variation compass, defined, 241
Vince, Rev. Samuel, FRS (1749-1821), 177, 178, 190, 195, 203, 205, 208; his *A complete system of astronomy*, 177
violin, Duncan Robertson's, 117-8
Visitation of Royal Observatory, 60, 167, 180, 198
Visitors of the Royal Observatory, 44, 60, 106, 116, 170, 222

Voltaire, François-Marie-Avouet de (1694-1778), 25

Waddington, Robert, 26, 28-34, 60, 216, 225-6, 245 (n.18), 247 (n.7); his *An Epitome of Theoretical and Practical Astronomy*, 42; his *The Sea Officer's Companion*, 42
Wales, William (1734?-98), 86, 110, 123, 126, 157, 175, 177, 180, 197
Walker, coachman, 194
Walpole, Frances Margaretta and Charlotte (NM's nieces), 205
Walsh, Eliza (NM's cousin) (1731-60), *see* Fowke
Walsh, Elizabeth (née Maskelyne) (NM's aunt) (1697-1734), 2
Walsh, John, HEIC, FRS (NM's cousin) (1726-95), 53, 113, 143, 252 (n.1)
Walsh, Joseph, HEIC (-1731)
Walsh, Joseph jr., (1722 <31)
Waring, Edward, DD, FRS (1734-98), 55, 76, 178
Warley East Indiaman, 188
Warrington, Earl of, *see* Booth
Warwick East Indiaman, 38-9, 42
Watson, Dr William the elder, FRS (1715-87), 147, 148
Watson, Dr William the younger, FRS (1744-1825?), 147
Webber, Capt., HEIC, 34
Wegg, Samuel, FRS, 234
Weld, Charles Richard (1813-69), 25
West, Benjamin, picture of death of Nelson, 208
West, James, PRS (1704?-72), 107, 249 (n.20)
Westcomb House, Greenwich, 100
Westminster Club Supper, 9
Westminster school, 3, 9, 214
White, Taylor, FRS, 44-5, 247 (n.12)
Williams, Walter, watchmaker, 82
Willoughby of Parham, Lord, FRS (-1765), 54, 247 (n.16)
Willoughby Fort, *see* Barbados
Wilson, Alexander, 137
Wilson, James, 54
Wilson, Patrick, 137
Windham, William, MP (1750-1810), 173-5
Windsor Castle, 150
Wingrave, Francis, publisher, 178
Winthrop, John, 34, 149
Witchell, George, FRS (1728-85), 74, 86, 121, 247 (n.7)
Wolfe, Gen. James (1727-59), 40
Wollaston, Rev. Francis, FRS (1731-1815), 54, 209
Wollaston, William Hyde, FRS, and Francis Hyde, FRS, 209
Woodwardian Professorship of Geology at Cambridge, 43

Woolf, Harry, 25
Wootton Bassett, 2, 164
Wray, Daniel, 22
Wreck Reef, 188
Wren, Sir Christopher, PRS (1632–1723), 63, 116
Wright, William, wine merchant, 230
Wright's Coffee House, 193

Yale University, 161

Yellow London lady, *see* Violin
York, Duke and Duchess of, 198

Zach, Franz Xaver, Baron von (1754–1802), 171, 174, 185–7, 256 (n.15)
zenith, defined, 241
zenith distance, 66; defined, 241
zenith sector, defined 241; *see also* Sisson